Nanomaterials under Extreme Conditions

A Systematic Approach to Designing and Applications

Editors

Manuel Ahumada Escandón

Centro de Nanotecnología Aplicada, Facultad de Ciencias, Ingeniería y Tecnología
Universidad Mayor, Santiago, RM
Escuela de Biotecnología, Facultad de Ciencias, Ingeniería y Tecnología
Universidad Mayor, Santiago, RM

and

María Belén Camarada Uribe

Departamento de Química Inorgánica, Facultad de Química y de Farmacia
Pontificia Universidad Católica de Chile, Santiago, Chile
Centro Investigación en Nanotecnología y Materiales Avanzados, CIEN-UC
Pontificia Universidad Católica de Chile, Santiago, Chile

CRC Press
Taylor & Francis Group
Boca Raton London New York

CRC Press is an imprint of the
Taylor & Francis Group, an **informa** business

A SCIENCE PUBLISHERS BOOK

Cover images taken from an OpenIA web page: Dall-E

First edition published 2023
by CRC Press
6000 Broken Sound Parkway NW, Suite 300, Boca Raton, FL 33487-2742

and by CRC Press
4 Park Square, Milton Park, Abingdon, Oxon, OX14 4RN

Library of Congress Cataloging-in-Publication Data (applied for)

ISBN: 978-0-367-46228-4 (hbk)
ISBN: 978-1-032-54812-8 (pbk)
ISBN: 978-1-003-02762-1 (ebk)

DOI: 10.1201/9781003027621

Typeset in Times New Roman
by Radiant Productions

Dedication

To my family, friends, colleagues, and especially those students who entrusted me with their future.

Manuel Ahumada Escandón

To Camilo and Sofie, my family, and all those friends who have supported me through this journey.

María Belén Camarada Uribe

Preface

Nanomaterials have become protagonists in different fields to support the finding of answers to the current needs of humankind. Such needs are transversal to the energy, environmental, industrial, and health sectors, where existing materials are not good enough either in terms of performance or efficiency. In this way, nanomaterials have allowed us to overcome many of the properties and flaws found in the same bulk materials and improve others. These small-size materials have shone because of their multiple physical-chemical properties, easiness, and variety of methodologies for their synthesis, and the almost unlimited composition combinations that allow the development of tailored nanomaterials, which can be fine-tuned to fulfill every type of need. As the knowledge of nanomaterials increases, the same happens to their applications, where even the new space race will play an important role. Therefore, nanotechnology has quickly become one of the most important industry sectors, with increasing revenues each year and an expected rise to the top places by 2030. However, still, there are important challenges that nanomaterials science must overcome, such as those related to the performance and efficiency of the material under different environmental conditions, particularly, under extreme conditions.

The design, preparation, and application of a nanomaterial must be realized considering the target or application itself and the environmental conditions in which it will be applied. which it will be applied. Therefore, this work seeks to complete that knowledge by promoting a critical view on the use of nanomaterials under extreme conditions found in our bodies, planet, and outer space. To this end, nanomaterials are covered from multiple points of view that allow the reader to get an enriching presentation of current knowledge on nanomaterials, limitations under extreme conditions, advancements within the field, and applications.

Redacted by nanomaterials science researchers, this book covers recent findings in nanomaterials under extreme conditions. Each chapter has been designed as a standalone document for lectors that are related to the topic or require a piece of specific information. However, newcomers interested in nanomaterials and nanotechnology are also welcome, as the book was designed as a complete fount of information to understand the nano concept and their applications under extreme conditions. During its first half, the historical conception, preparation, physical-chemical properties, and characterization techniques are explored, allowing the reader to understand the importance of designing nanomaterials. In its second half, the use of nanomaterials under extreme conditions is delved into, offering an excellent material source for understanding and discussion about the relevance of the environmental conditions in their preparation.

Dr. Manuel Ahumada Escandón
Director, Center for Applied Nanotechnology, Faculty of Sciences, Engineering, and Technology, Universidad Mayor, Santiago, Chile

Dr. María Belén Camarada Uribe
Assistant Professor, Department of Inorganic Chemistry, Faculty of Chemistry and Pharmacy, and Centro Investigación en Nanotecnología y Materiales Avanzados, CIEN-UC, Pontificia Universidad Católica de Chile, Santiago, Chile

Acknowledgments

Manuel Ahumada Escandón, Ph.D., thanks the financial support of ANID-FONDECYT grant #11180616, and the Center for Applied Nanotechnology and the Universidad Mayor for their continued support.

María Belén Camarada, Ph.D., thanks the financial support of ANID-FONDECYT grant #1180023, and the Faculty of Chemistry and Pharmacy of Pontificia Universidad Católica de Chile, especially to the members of the Laboratory of Functional Materials.

Contents

Chapter **1**

Nanomaterials under Extreme Conditions
Why Does It Matter?

María Belén Camarada[1,2,*] and *Manuel Ahumada*[3,4,*]

1. Introduction

There is no doubt that the development of new materials and their applications have marked most of humanity's eras, going from the stone, copper, bronze, iron, industrial, atomic, space, and information ages, to name a few (Valiulis 2014). In every one of them, more significant milestones related to materials development have helped push humanity to the next frontiers.

Together with the development of new materials, there have also been opportunities to promote a better understanding of their phenomena, such as the case of nanomaterials. There are reports of nanoparticles in materials as old as the 4th century with the Lycurgus cup, in the stained-glass windows in Europe cathedrals during the 6th–15th centuries, or in the Damascus saber blades (Bayda et al. 2019). However, it was not until the 20th century that nanoscience saw its first installments as a science field. Landmarks such as the discovery of colloidal gold, calculation of the glucose molecule size, invention of the field emission microscope, and the anteroom for the historic Feynman's talk. "There's plenty of room at the bottom" in 1959 was the first mention of the possibility of using materials at the atomic scale (Feynman 1959). Further iconic discoveries and inventions followed Feynman's

[1] Laboratorio de Materiales Funcionales, Departamento de Química Inorgánica, Facultad de Química y de Farmacia, Pontificia Universidad Católica de Chile, Santiago, 7820436, Chile.
[2] Centro Investigación en Nanotecnología y Materiales Avanzados, CIEN-UC, Pontificia Universidad Católica de Chile, Santiago, Chile.
[3] Centro de Nanotecnología Aplicada, Facultad de Ciencias, Ingeniería y Tecnología, Universidad Mayor, Camino La Pirámide 5750, Huechuraba, Santiago, RM.
[4] Escuela de Biotecnología, Facultad de Ciencias, Ingeniería y Tecnología, Universidad Mayor, Camino La Pirámide 5750, Huechuraba, Santiago, RM.
* Corresponding authors: mbcamara@uc.cl; manuel.ahumada@umayor.cl

talk until 1974, when the term "nanotechnology" was coined by Norio Taniguchi, defining it as "the process of separation, consolidation, and deformation of materials by one atom or one molecule" (Sandhu 2006). By the beginning of the 90s, atomic manipulation and the discovery of carbon nanotubes triggered the interest of several governments to explore nanotechnology as a potential industry to generate new income for the countries. In this way, by the year 2000, the United States of America was the first to establish nanotechnology as a priority development area by creating the National Nanotechnology Initiative (NNI). Immediately, other countries such as Canada, Japan, the European Union, and Brazil, among others, followed the same or similar pathways (Lazurko et al. 2019). Nowadays, nanotechnology has reached a top position among the most critical industries. It applies to every field involving, but not limited to, energy, environment, data, space, engineering, construction, agronomy, and medicine. The following sections of this chapter explore the term nanomaterial and briefly describe the synthetic approaches, physico-chemical properties, and typical characterizations. Further, the concept of the extreme condition is established with the applications of nanomaterials under such conditions.

2. Nanomaterials

Nanomaterials, as systems, are invisible to the naked eye, given that their sizes are smaller than a cell, being between the sizes of molecules and viruses, as depicted in Figure 1. Their classical definition described them as molecular or atomic arranges, intentionally prepared, with structure sizes ranging between 1–100 nm in at least one of its dimensions (Mansoori and Soelaiman 2005). However, it must be considered that nanomaterial definition changes from country to country, implying differences in established laws and norms (Lazurko et al. 2019).

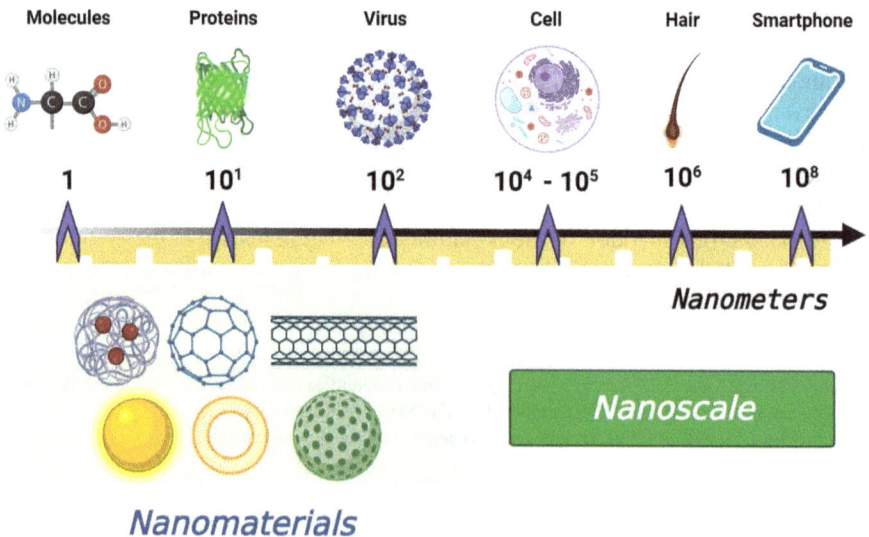

Figure 1. Depiction of the nanoscale. Nanomaterials' size range 1–100 nm, i.e., between individual molecules and virus sizes.

Of course, nanomaterials are diverse, and in the literature, several types of popular classifications can be found depending on their composition. For instance, they can be classified as organic or inorganic, depending on the material source used for their preparation. Also, natural, or synthetic, depending on if the used molecules have a natural or synthetic origin/source. Other classifications order them within subcategories related to a typical composition (e.g., carbon-based, polymer-based, and metal-based) or parameters (e.g., electroconductivity, optically active, and biocompatibility). Although their possible classifications are based on composition, it should always be considered that the composition is not the only characteristic of nanomaterials, as they also can be described as a function of their shape or morphology (spherical, planar, rod, etc.), or by their structure (solid, hollow, multiwalled, etc.) (Buzea and Pacheco 2017).

3. Synthesis Pathways

There are hundreds of methodologies reporting nanomaterials synthesis. Nonetheless, they can be classified into two broad categories: top-down and bottom-up pathways (Figure 2) (Biswas et al. 2012). The first one produces the nanomaterials starting from bulk material and "reducing it" to a nanoparticulated state; this process is generally reached by using physical methods of high energy, such as high-energy lasers, ionization, and mechanics or thermal energy, to name a few (Iqbal et al. 2012). On the other hand, bottom-up methods use the atoms (from salts, for example) or molecules, and by chemical or physical reaction(s), they can grow to promote nanomaterial formation (Needham et al. 2016). These pathways can be carried out

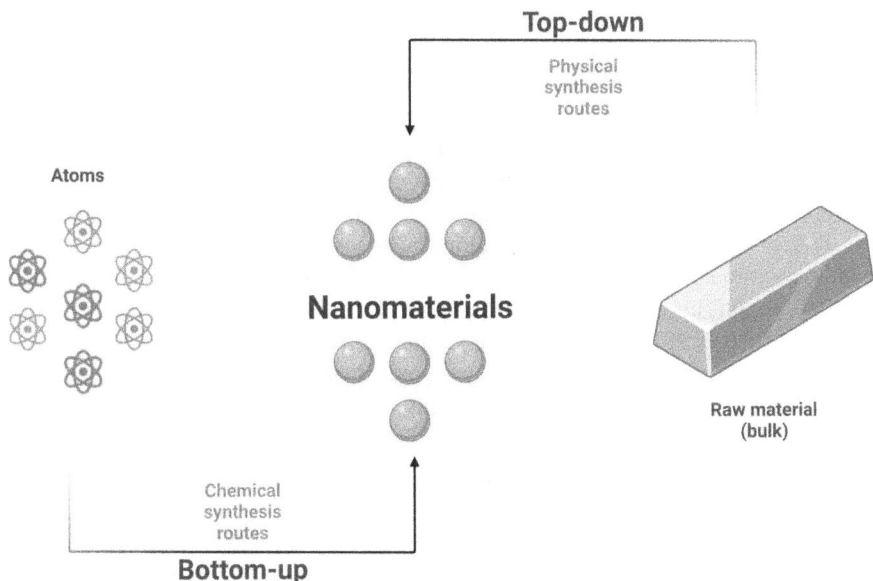

Figure 2. Schematic representation of the bottom-up and top-down pathways for nanomaterials preparation.

in gas, liquid, and/or solid phases, and also using biological sources (Luechinger et al. 2010). Therefore, while in one case, it is needed to "disassemble the material," in the other, the "assembly of the components" is required, respectively (Narayanan and Sakthivel 2010).

Among the most usual methods within the top-down pathway, it is possible to find:

(i) Thermal evaporation involves heating the bulk material until it reaches material evaporation. It is carried out in a vacuum chamber where the vapor is condensed over a cold layer, requiring precise control of the experimental growth conditions to avoid promoting morphology modifications of the deposited layer (Biswas et al. 2012).

(ii) Chemical vapor deposition (CVD): It consists of the decomposition of one or several volatile compounds within a vacuum chamber (reactor) and posterior deposition over a solid surface, giving place to the formation of a monolayer or film of the nanomaterial (Elliott et al. 2013).

(iii) Gaseous clusters: They consist of the use of high-power pulsed lasers to produce vapor of metallic atoms, which are further carried with inert gas, and posteriorly deposited on a monocrystalline oxide or another substrate under conditions of ultra-high vacuum (Arole and Munde 2014).

In contrast, the bottom-up pathway is commonly characterized by chemical procedures. There are many described methodologies about this method, such as redox reactions, co-precipitation, hydrothermal reactions, polymerization, and photochemical processes, to name a few (Abid et al. 2022). Let's consider this last one as an example of metal nanoparticle preparation; here, the metal ions reduction is carried out by free radicals generated by light irradiation. In this way, the free radicals act as reducing agents, highly active, that promote the reduction of the metal ion to its metallic form, and by posterior support of a capping agent, the formation of the nanoparticle is reached (Chen et al. 2015).

It could be appreciable that the top-down approach methodologies will require more complex and expensive instrumentations and conditions. On the other hand, the bottom-up pathway does not require such instrumentation, but experimental conditions should always be kept under control. The aforementioned has prompted that the bottom-up approaches have a higher prevalence within the literature and industry (Biswas et al. 2012). Of course, the methods are far more complex than this brief description, and while in this book the reader can revise some of those methodologies, we recommend further lectures in related manuscripts.

4. Physicochemical Properties

The prepared nanomaterials will have physicochemical properties that differ from those observed in their respective bulk materials, where common properties such as optical, thermal, conductive, mechanical, and biological properties, among others, can be fine-tuned to obtain desired results (Khan et al. 2019). Next, some of these properties are briefly explored:

(i) Optical and electrical properties of conductive nanomaterials: In these cases, the properties are observed because of the interaction of the nanomaterial's surface electrons with the incident electromagnetic wave (e.g., natural light, diodes, lasers, etc.), generating a quantic effect that redistributes the surface electrons (polarization), also known as localized surface plasmon resonance (Eustis and El-Sayed 2006). The coupling of the plasmon of a nanomaterial with an incident photon can improve several useful optical phenomena, such as nuclear resonance scattering, Raman dispersion, and luminescence processes modulation. Due to these unique features, the plasmonic structures of different sizes and shapes have increased in popularity in cancer diagnostic, photothermal therapy, and live bioimaging and detection, to name some (Madden et al. 2015).

(ii) Mechanical properties: These are relative to the size. At the nanoscale, there is an atomic structure modification of the nanocrystals, making them more resistant with superior mechanical properties than their bulk counterpart. In this sense, within the nanosized range, a material increases its hardness and resistance, following, generally, an inverse proportion concerning its size (Khan et al. 2019). Another relevant property of nanomaterials is their higher deformation capacity to the traction before the breakpoint. As such, the nanomaterials have the quality to support elevated external tensions without promoting structural dislocation of the nanocrystal and without fissures or fractures (Guo et al. 2013).

(iii) Surface-to-volume ratio: One of the most relevant features of nanomaterials, implying that the available surface area is bigger than their volume. For example, a single perfectly spherical nanoparticle of 10 nm in diameter will have a volume value of 5.25×10^{-25} m^3, while its area value will be close to 3.14×10^{-16} m^2. If we consider a mol of nanoparticles, those values elevate to 0.32 m^3 and 189×10^6 m^2, respectively. These numbers exemplify that while nanomaterials can occupy small volumes, they have a high available surface that can be used as an active site for different applications. As most of the atoms occupy internal layers in the bulk material, in the nanomaterials, a significant portion of the atoms forms part of the surface (Yokoyama et al. 2008). The atoms on the surface of the nanomaterials are unstable, have higher energic levels, and the forces by which they are attracted to the core of the material are weak (Phan and Haes 2019). This surface capacity favors the interfacial energy increment, reactivity, absorption efficiency, and capacity to make it functional with molecules of interest (Sperling and Parak 2010).

In Chapter 2, there is an in-detail description of the physical-chemical properties of nanomaterials. Further, in every chapter, these properties are explored in-depth to rationalize the implementation of nanomaterials under extreme conditions.

5. Characterization Techniques

Essential after synthesizing any nanomaterial is their characterization, which can give information about size, morphology, shape, composition, stability, and interactions, among others. To fulfill such an endeavor, it is possible to utilize routinary assays that go from naked-eye observations to more complex measurements that require more specialized equipment such as transmission electron microscopy (TEM) or X-ray photoelectron spectroscopy (XPS), to name a few. Table 1 shows a summary of some of the most common characterization techniques. Nonetheless, in Chapter 3, it is possible to find an in-depth description of these and other additional methods.

6. Extreme Conditions

Nanomaterials have been utilized for different applications and tested for several parameters; however, those are commonly referred to as standard conditions. While these have helped solve problems in various fields, the need to continue improving current technologies and further expanding current technological frontiers has prompted the need to develop and test the nanomaterials under the harshest conditions. When referring to extreme conditions, these are considered environmental conditions where the working parameters are outside the standard norm (e.g., pressure, temperature, gravity, among others). If these are referred to the humans, they should be interpreted as the environmental conditions where to live is extremely limited or impossible. In the case of nanomaterials, extreme conditions are those where the materials can suffer structural or compositional modifications that can deny their potential application, affect concomitant structures/materials, and ultimately impede new nanomaterials' development. Some examples of extreme conditions for nanomaterials are given next.

High temperatures: The condition in which the temperature exceeds 600°C approximately. Many crucial industrial and aerospace processes are carried on at this temperature and above, such as catalytic combustion, steam reforming, and production, among others. The thermal stability and energy release that the nanomaterials can provide are essential to evaluate and improve (Zarur and Ying 2000).

High pressure: The condition in which the pressure is above the GPa range (Jayaraman 1986). Under this condition, pressure-induced phase transitions in the nanomaterial can impact their surface (e.g., topology) and mechanical (e.g., stress) properties and performance.

Corrosiveness: The conditions in which the environment promotes an irreversible interfacial reaction on any nanomaterial. Under corrosive environments, nanomaterials can undergo several modifications in their properties that will ultimately impact their performance and economical cost (Groysman 2010).

Biological: The condition in which the nanomaterial is within or surrounded by an organic living environment. Under these conditions, the material can interact with

Table 1. Standard techniques commonly employed to characterize nanomaterials.

Technique	Description	Application examples	Reference
UV-Vis spectroscopy	Determine the absorption/ transmittance of light by a sample	Plasmon and turbidimetry determinations	(Okitsu 2013)
Fluorescence spectroscopy	Determine the excitation and light emission by a sample	Photoluminescence and quenching determinations	(Cao et al. 2012)
Fourier transform-Infrared spectroscopy (FT-IR)	Establish the presence of functional groups within the molecular composition	Composition and functionalization characterizations	(Baudot et al. 2010)
Raman spectroscopy	It gives a structural fingerprint through analysis of the vibrational modes	Molecular identification and surface-enhanced Raman scattering (SERS)	(Gouadec and Colomban 2007)
X-ray photoelectron spectroscopy	Determine the elements and their chemical state present on the surface of a nanomaterial	Quantitative elemental analysis and chemical characterization of nanomaterial's surface	(Baer and Engelhard 2010)
X-ray diffraction (XRD)	Provide information on crystal structure, crystallite size, and strain	Crystallinity and structural characterization of solid samples	(Giannini et al. 2016)
Atomic force microscopy (AFM)	Non-optical microscopy that makes use of a non-destructive scanning probe for surface properties' evaluation	Topographic imaging and surface force measurements	(Zhong and Yan 2016)
Scanning electron microscopy (SEM)	Focused electron beam that scans the surface	Surface observation and analysis of micro- and nanomaterials	(Akhtar et al. 2018)
Transmission electron microscopy (TEM)	Focused electron beam that transmits through the sample	Observation and analysis of nanomaterials	(Kumar 2013)
Differential scanning calorimetry (DSC)	Detects the possible exothermic and endothermic processes of a sample	Phase transition and enthalpy determination	(Bannov et al. 2020)
Thermal gravimetric analysis (TGA)	Measures the change in mass of a sample due to the temperature variations over time	Sample decomposition, oxidation, and evaporation temperature	(Loganathan et al. 2017)
Dynamic light scattering (DLS)	Evaluate the changes in the Brownian motion of nanomaterials by light dispersion	Size and polydispersity determination	(Bhattacharjee 2016)
ζ potential	Measures the electrophoretic mobility of a sample	Surface charge potential as a parameter of colloidal stability	(Varenne et al. 2015)
Nanoparticle tracking analysis (NTA)	Measures the changes in the visualized Brownian motion of nanoparticles	Size and concentration determination	(Patois et al. 2012)

cells, proteins, lipids, and DNA, which could positively or negatively impact the living organism, i.e., affect human health (Alarcon and Ahumada 2019).

Microgravity: The condition in which the gravity is close to zero. Under such conditions, the objects experience weightlessness, but nanoscale processes such as crystal growth, capillary driven flow, and diffuse transport can easily influence the nanomaterial's properties and effectiveness (Mohr et al. 2019).

Along with this book, these extreme conditions will be further described and explored from the nanomaterial's perspective.

Outlook

This chapter exposed a brief overview of the topics this book covers. Nanomaterials have significantly impacted humankind's development. They have helped the development of impossible solutions in almost every field related to worldwide needs in industry, energy, environment, and health, among others. Researchers have been able to tailor the nanomaterials to fine-tune almost every need to reach such advancements, following one of the two main synthesis pathways, "bottom-up" and "top-down." The posterior physical-chemical characterization of the nanomaterial will provide evidence of its characteristics and properties, such as optical, conductivity, thermal, active surface, and mechanical, to name a few. Despite the advancement in solving the current needs, still some places are hard to reach or explore due to the environmental conditions, where existing materials cannot tolerate them. Then, developing new nanomaterials that can work under extreme conditions has and will play a principal role in developing new technologies and solutions.

Acknowledgments

M.A. thanks FONDECYT grant "Iniciación a la Investigación" #11180616, and Universidad Mayor grant "Proyecto Iniciación-2019091". M.B.C. thanks FONDECYT Regular #1180023. M.A and M.B.C. thanks all the authors who have contributed to develop this book. Figures were created with BioRender.com.

References

Abid, Namra, Aqib Muhammad Khan, Sara Shujait, Kainat Chaudhary, Muhammad Ikram, Muhammad Imran, Junaid Haider, Maaz Khan, Qasim Khan and Muhammad Maqbool. (2022). Synthesis of nanomaterials using various top-down and bottom-up approaches, influencing factors, advantages, and disadvantages: A review. *Advances in Colloid and Interface Science*, 300: 102597.

Akhtar, Kalsoom, Shahid Ali Khan, Sher Bahadar Khan and Abdullah M. Asiri. (2018). Scanning electron microscopy: Principle and applications in nanomaterials characterization. In *Handbook of Materials Characterization* (Springer).

Alarcon, Emilio I. and Manuel Ahumada. (2019). *Nanoengineering Materials for Biomedical Uses* (Springer Nature: Switzerland).

Arole, Dr. V.M. and Prof. S.V. Munde. (2014). Fabrication of nanomaterials by top-down and bottom-up approaches—An overview. *Journal of Advances in Applied Sciences and Technology*, 1: 89–93.

Baer, Donald R. and Mark H. Engelhard. (2010). XPS analysis of nanostructured materials and biological surfaces. *Journal of Electron Spectroscopy and Related Phenomena*, 178: 415–32.

Bannov, Alexander G., Maxim V. Popov and Pavel B. Kurmashov. (2020). Thermal analysis of carbon nanomaterials: Advantages and problems of interpretation. *Journal of Thermal Analysis and Calorimetry*, 142: 349–70.

Baudot, Charles, Cher Ming Tan and Jeng Chien Kong. (2010). FTIR spectroscopy as a tool for nano-material characterization. *Infrared Physics & Technology*, 53: 434–38.

Bayda, Samer, Muhammad Adeel, Tiziano Tuccinardi, Marco Cordani and Flavio Rizzolio. (2019). The history of nanoscience and nanotechnology: From chemical-physical applications to nanomedicine. *Molecules (Basel, Switzerland)*, 25: 112.

Bhattacharjee, Sourav. (2016). DLS and zeta potential–what they are and what they are not? *Journal of Controlled Release*, 235: 337–51.

Biswas, Abhijit, Ilker S. Bayer, Alexandru S. Biris, Tao Wang, Enkeleda Dervishi and Franz Faupel. (2012). Advances in top–down and bottom–up surface nanofabrication: Techniques, applications & future prospects. *Advances in Colloid and Interface Science*, 170: 2–27.

Buzea, Cristina and Ivan Pacheco. (2017). Nanomaterials and their Classification. *In*: Ashutosh Kumar Shukla (ed.). *EMR/ESR/EPR Spectroscopy for Characterization of Nanomaterials* (Springer India: New Delhi).

Cao, Bingqiang, Haibo Gong, Haibo Zeng, W.P. Cai, Kaushal Kumar, Luigi Sanguigno, Filippo Causa, Paolo Antonio Netti and Yashashchandra Dwivedi. (2012). Photoluminiscence/Fluorescence spectroscopic technique for nanomaterials characterization. *Nanomaterials: Processing and Characterization with Lasers, Weinheim, Wiley*, 597–616.

Chen, Qiang, Junshuai Li and Yongfeng Li. (2015). A review of plasma–liquid interactions for nanomaterial synthesis. *Journal of Physics D: Applied Physics*, 48: 424005.

Elliott, James A., Yasushi Shibuta, Hakim Amara, Christophe Bichara and Erik C. Neyts. (2013). Atomistic modelling of CVD synthesis of carbon nanotubes and graphene. *Nanoscale*, 5: 6662–76.

Eustis, Susie and Mostafa A. El-Sayed. (2006). Why gold nanoparticles are more precious than pretty gold: Noble metal surface plasmon resonance and its enhancement of the radiative and nonradiative properties of nanocrystals of different shapes. *Chemical Society Reviews*, 35: 209–17.

Feynman, Richard P. (1959). There's plenty of room at the bottom. *Engineering and Science*, 23.

Giannini, Cinzia, Massimo Ladisa, Davide Altamura, Dritan Siliqi, Teresa Sibillano and Liberato De Caro. (2016). X-ray diffraction: A powerful technique for the multiple-length-scale structural analysis of nanomaterials. *Crystals*, 6: 87.

Gouadec, Gwénaël and Philippe Colomban. (2007). Raman Spectroscopy of nanomaterials: How spectra relate to disorder, particle size and mechanical properties. *Progress in Crystal Growth and Characterization of Materials*, 53: 1–56.

Groysman, A. (2010). Corrosion mechanism and corrosion factors. In *Corrosion for Everybody* (Springer Netherlands: Dordrecht).

Guo, Dan, Guoxin Xie and Jianbin Luo. (2013). Mechanical properties of nanoparticles: Basics and applications. *Journal of Physics D: Applied Physics*, 47: 013001.

Iqbal, Parvez, Jon A. Preece and Paula M. Mendes. (2012). Nanotechnology: The "Top-Down" and "Bottom-Up" approaches. *In Supramolecular Chemistry* (John Wiley & Sons, Ltd).

Jayaraman, Aiyasami. (1986). Ultrahigh pressures. *Review of Scientific Instruments*, 57: 1013–31.

Khan, Ibrahim, Khalid Saeed and Idrees Khan. (2019). Nanoparticles: Properties, applications and toxicities. *Arabian Journal of Chemistry*, 12: 908–31.

Kumar, Challa S.S.R. (2013). *Transmission Electron Microscopy Characterization of Nanomaterials* (Springer Science & Business Media).

Lazurko, Caitlin, Manuel Ahumada, Emilio I. Alarcon and Erik Jacques. (2019). Regulatory normative of nanomaterials for their use in biomedicine. *In*: Emilio I. Alarcon and Manuel Ahumada (eds.). *Nanoengineering Materials for Biomedical Uses* (Springer International Publishing: Cham).

Loganathan, Sravanthi, Ravi Babu Valapa, Raghvendra Kumar Mishra, Pugazhenthi, G. and Sabu Thomas. (2017). Thermogravimetric analysis for characterization of nanomaterials. In *Thermal and Rheological Measurement Techniques for Nanomaterials Characterization* (Elsevier).

Luechinger, Norman A., Robert N. Grass, Evagelos K. Athanassiou and Wendelin J. Stark. (2010). Bottom-up fabrication of metal/metal nanocomposites from nanoparticles of immiscible metals. *Chemistry of Materials*, 22: 155–60.

Madden, Olena, Michael Daragh Naughton, Siobhan Moane and Patrick G. Murray. (2015). Mycofabrication of common plasmonic colloids, theoretical considerations, mechanism and potential applications. *Advances in Colloid and Interface Science*, 225: 37–52.

Mansoori, G. Ali and Fauzi Soelaiman, T.A. (2005). *Nanotechnology—An Introduction for the Standards Community* (ASTM International).

Mohr, Markus, Rainer K. Wunderlich, Kai Zweiacker, Silke Prades-Rödel, Romuald Sauget, Andreas Blatter, Roland Logé, Alex Dommann, Antonia Neels, William L. Johnson and Hans-Jörg Fecht. (2019). Surface tension and viscosity of liquid Pd(43)Cu(27)Ni(10)P(20) measured in a levitation device under microgravity. *NPJ Microgravity*, 5: 4–4.

Narayanan, K.B. and Sakthivel, N. (2010). Biological synthesis of metal nanoparticles by microbes. *Adv. Colloid Interface Sci.*, 156: 1–13.

Needham, D., Arslanagic, A., Glud, K., Hervella, P., Karimi, L., Høeilund-Carlsen, P.F. Kinoshita, K., Mollenhauer, J., Parra, E., Utoft, A. and Walke, P. (2016). Bottom up design of nanoparticles for anti-cancer diapeutics: "put the drug in the cancer's food. *J. Drug Target*, 24: 836–56.

Okitsu, Kenji. (2013). UV-Vis spectroscopy for characterization of metal nanoparticles formed from reduction of metal ions during ultrasonic irradiation. *In*: Challa Kumar (ed.). *UV-VIS and Photoluminescence Spectroscopy for Nanomaterials Characterization* (Springer Berlin Heidelberg: Berlin, Heidelberg).

Patois, E., Capelle, M.A.H., Palais, C., Gurny, R. and Arvinte, T. (2012). Evaluation of nanoparticle tracking analysis (NTA) in the characterization of therapeutic antibodies and seasonal influenza vaccines: Pros and cons. *Journal of Drug Delivery Science and Technology*, 22: 427–33.

Phan, H.T. and Haes, A.J. (2019). What does nanoparticle stability mean? *The Journal of Physical Chemistry. C, Nanomaterials and Interfaces*, 123: 16495–507.

Sandhu, Adarsh. (2006). Who invented nano? *Nat. Nanotechnol.*, 1: 87–87.

Sperling, R.A. and Parak, W.J. (2010). Surface modification, functionalization and bioconjugation of colloidal inorganic nanoparticles. *Philosophical Transactions of the Royal Society A: Mathematical, Physical and Engineering Sciences*, 368: 1333–83.

Valiulis, Algirdas Vaclovas. (2014). The place of Technology in history. *In*: A History of Materials and Technologies Development (VGTU press TECHNIKA).

Varenne, Fanny, Jérémie Botton, Claire Merlet, Jean-Jacques Vachon, Sandrine Geiger, Ingrid C. Infante, Mohamed M. Chehimi and Christine Vauthier. (2015). Standardization and validation of a protocol of zeta potential measurements by electrophoretic light scattering for nanomaterial characterization. *Colloids and Surfaces A: Physicochemical and Engineering Aspects*, 486: 218–31.

Yokoyama, Toyokazu, Hiroaki Masuda, Michitaka Suzuki, Kensei Ehara, Kiyoshi Nogi, Masayoshi Fuji, Takehisa Fukui, Hisao Suzuki, Junichi Tatami and Kazuyuki Hayashi. (2008). Basic properties and measuring methods of nanoparticles. In *Nanoparticle Technology Handbook* (Elsevier).

Zarur, Andrey J. and Jackie Y. Ying. (2000). Reverse microemulsion synthesis of nanostructured complex oxides for catalytic combustion. *Nature*, 403: 65–67.

Zhong, Jian and Juan Yan. (2016). Seeing is believing: Atomic force microscopy imaging for nanomaterial research. *RSC Advances*, 6: 1103–21.

Chapter **2**

Nanomaterials
Physical and Chemical Properties

Natalia L. Pacioni[1,2,*] *and M. Andrea Molina Torres*[1,3]

1. Introduction

Nanomaterials (NM) comprise a vast class of compounds whose size is typically between 1 to 100 nm, at least in one dimension. As a consequence of this nanometric size, the physical and chemical properties are remarkably different from those observed in the bulk material. For example, in the 1970s, it was discovered that metals like copper and silver, well-known for their electrical and thermal conductivities, in the nanoscale would lose these properties becoming nonconductor and nonthermal (Shi et al. 2015). Therefore, both size and modification of properties are generally used to define NM (Auffan et al. 2009). Besides, using additional criteria, NM can be classified according to chemical composition, morphology, and dimensionality as showed in Table 1 (Buzea et al. 2007).

The high applicability of NM is closely related to their physical and chemical properties. Consequently, a deep understanding of them is relevant to predict their behavior under different experimental conditions or natural sceneries. In the following sections, different NM properties and their dependence on several NM characteristics are described.

[1] Universidad Nacional de Córdoba, Facultad de Ciencias Químicas, Departamento de Química Orgánica. Haya de la Torre y Medina Allende s/n, X5000HUA, Ciudad Universitaria, Córdoba, Argentina.
[2] Consejo Nacional de Investigaciones Científicas y Técnicas (CONICET), INFIQC, Haya de la Torre y Medina Allende s/n, X5000HUA, Ciudad Universitaria, Córdoba, Argentina.
[3] Consejo Nacional de Investigaciones Científicas y Técnicas (CONICET), IPQA, Haya de la Torre y Medina Allende s/n, X5000HUA, Ciudad Universitaria, Córdoba, Argentina.
* Corresponding author: n.lpacioni@unc.edu.ar

Table 1. Classification of nanomaterials.

Criteria	Class	Sub-class	Examples
Chemical composition	Inorganic NM	metal nanoparticles	• AgNP • AuNP • CuNP
		metal oxide nanoparticles	• CeO • TiO2 • ZnO • CuO
		magnetic nanoparticles	• Fe_3O_4 • γ-Fe_2O_3
		quantum-dots	• CdSe • CdS • ZnSe
	Organic NM	dendrimers	
		liposomes	
	Carbon-based NM	fullerenes	
		carbon-dots	
		graphene	
		carbon nanotubes	
Morphology	Low-aspect ratio	nanospheres	
		nanocubes	
		nanorods	
		nanopyramids	
	High-aspect ratio	nanowires	
		nanobelts	
Dimensionality	1D	thin films	
		surface coatings	
	2D	- free high-aspect ratio nanowires	
		- thick membranes with nanopores	
		- fixed long nanostructures on a support	
	3D	- membranes with nanopores	
		- fixed small nanostructures on a support	
		- free small aspect ratio nanoparticles	

2. Physical and Chemical Properties

2.1 Electrical Properties

There are three categories of materials based on their electrical properties: conductors, semiconductors, and insulators. The difference in the energy between the valence (VB) and the conduction (CB) bands is called *bandgap energy* (E_g). The ability to fill the CB with electrons from the VB, and the value of E_g determine if the material is conductive, semiconductor or insulator (Figure 1).

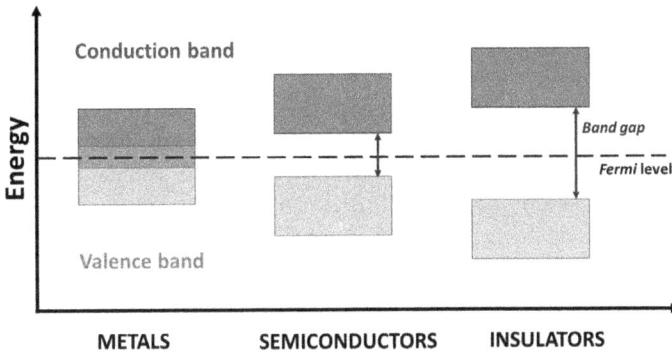

Figure 1. Representation of the bandgap energy diagram for metals, semiconductors, and insulators.

In **conductive** materials, such as metals, the VB and the CB overlap, so the E_g value is small enough for thermal energy to be sufficient to stimulate the electrons to move to the CB.

In **semiconductors**, the E_g value is a few electrovolt (≤ 3.5 eV). If the application of a voltage exceeds the E_g, the electrons in the VB promote to the CB, forming electron-hole pairs called *excitons*.

In contrast, **insulators** have a very large Eg value (> 3.5 eV) that requires enormous amounts of voltage to exceed that limit. Therefore, these materials do not conduct electricity.

As the particle size decreases to nanometric values, the effect of quantum confinement causes the *bandgap energy* to increase, and consequently, some metal oxides start to behave as semiconductors; for example, this effect is observed in quantum-dots (Zhang 2009).

The electrical properties of some nanomaterials are related to their unique structures and exhibit electrical properties that are exceptional. For example, single-wall CNTs with tooth-shaped structures are conductors, whereas the palm-shaped ones have a semiconductor behavior (Shi et al. 2015).

2.2 Optical Properties

Light can interact with matter to provoke different well-known phenomena. For example, the incident energy can promote electrons from the ground electronic state to the excited singlet electronic states (*UV-visible absorption*). Then, the molecule can return to the ground state by, for example, emitting light (*fluorescence*). These photophysical phenomena as well as others (*infrared absorption, Raman scattering, phosphorescence*) depend on many variables, and several parameters are used to define their photophysical properties, such as molar absorption coefficient, ε, and fluorescence quantum yield, Φ.

In the case of NM, the optical properties depend on variables like size, shape, surface characteristics, and their interaction with the environment around it (ligands, solvents) or with other nanostructures. For example, when the particle size decreases, a blue shift is observed in metal nanoparticles (MNP) absorbance or the quantum dots photoluminescence spectra (Zhang 2009).

Noble MNP, such as AuNP, AgNP, and CuNP, exhibit an intense extinction band in the visible region due to the localized surface plasmon resonances (SPR, Figure 2), and the coherent oscillations of the electrons of a metal in the vicinity of a metal-dielectric interface (Coronado et al. 2011). Thus, when the frequency of the incident photon matches the SPR frequency of the metal, an absorption band shows up in the spectrum. The resonance frequencies of SRP for some metals, such as Pb, In, Hg, Sn, and Cd, are found in the UV region of the electromagnetic spectrum, then these nanoparticles do not exhibit intense color effects (Ghosh and Pal 2007).

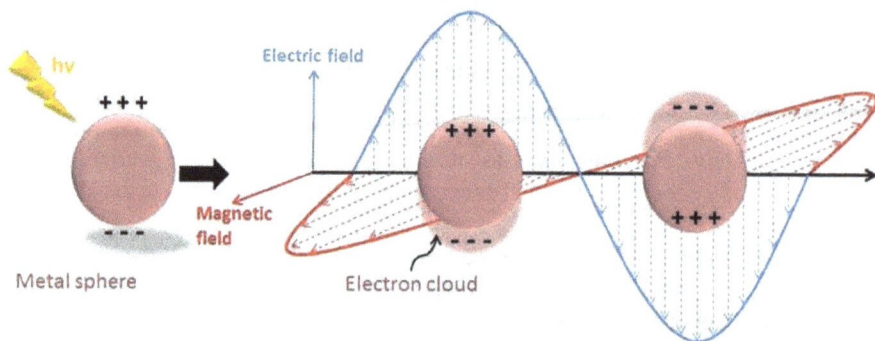

Figure 2. Schematic representation of localized SPR for a metal nanosphere. Reprinted with permission from (Peiris et al. 2016). Copyright® Royal Society of Chemistry.

Four factors mainly determine the oscillation frequency: the electrons density, the effective electron mass, the shape, and the size of the charge distribution. The extinction band is a combination of absorption and scattering phenomena and is thus highly dependent on the particle size (Amendola et al. 2017, Lakowicz 2005). For example, for AuNP, if $d \ll \lambda$; where d is the size, and λ is the wavelength of incident electromagnetic radiation, absorption prevails over scattering. However, light dispersion and absorption depend on the 6th and the 3rd power of a particle's size, respectively. Therefore, the effects become comparable at $d \approx 50$ nm. For larger particles, scattering dominates over absorption (Amendola et al. 2017, Valcárcel et al. 2014). Furthermore, some parameters like the presence of a supporting substrate or ligand, the solvent, and the electromagnetic interparticle interactions must be considered when analyzing the optical properties of MNP (Ghosh and Pal 2007). The influence of different factors on the SPR of MNP is described as follows.

2.2.1 *Factors Affecting the Surface Plasmon Resonance*

As mentioned above, the size of MNP critically influences the SPR band position in the extinction spectrum. Size provokes two distinctive effects named as extrinsic and intrinsic.

The *extrinsic size* influence corresponds to a retardation effect due to the excitation of multipolar plasmon modes when the nanoparticle size increases with respect to the incident wavelength. In other words, when the electric field distribution along the nanoparticle is non-uniform, multipolar plasmons oscillations are excited, and then, a broadening and redshift of the SPR band appears in the absorption spectrum for particles whose size parameter, defined as the ratio between d and l, approximates to 1 (Amendola et al. 2017).

The *intrinsic size effects* are responsible for the SPR damping observed when the particle size decreases below 30 nm (Figure 3) because of the additional contributions to the free-electron relaxation rate, which modifies the optical constant of the metal. It is possible to achieve a near-complete SPR quenching when the size is smaller than 2 nm (Amendola et al. 2017).

The width, position, and number of SPR bands are also influenced by the particle shape. For example, spherical AuNP sizes between 2–50 nm have an SPR band centered around 520 nm, unlike nanorods (AuNR) that have two SPR bands. The first one, at high energies, corresponds to the oscillation of transverse electrons, while the second band located at lower energies belongs to the oscillation of longitudinal electrons (Link and El-Sayed 1999). The aspect ratio (AR) in nanorods also influences the SPR longitudinal wavelength position, provoking a redshift when AR increases (Figure 4).

The physical-chemical environment around the nanoparticles also influences the SPR in different ways. The main mechanism involved corresponds to the chemical interface damping (CID). This mechanism consists of the direct transfer energy of interfacial "hot" electrons from a metal to a molecule or material in the vicinity of the metal surface (Lee et al. 2019). The phenomenon can take place by chemisorption, for example, when molecules attach to the surface of nanoparticles by thiol bonds or by physisorption, such as in the case of citrate-stabilized nanoparticles.

The presence of the adsorbate bond to the metal surface offers new routes of energy relaxation for the excited electrons and phonons in the metal. Then, the CID phenomenon provokes the broadening and redshift of the SPR band (Valcárcel et al. 2014) due to the several couplings produced between the adsorbate lowest unoccupied molecular orbital (LUMO) and the free electrons in the metal conduction (Amendola et al. 2017).

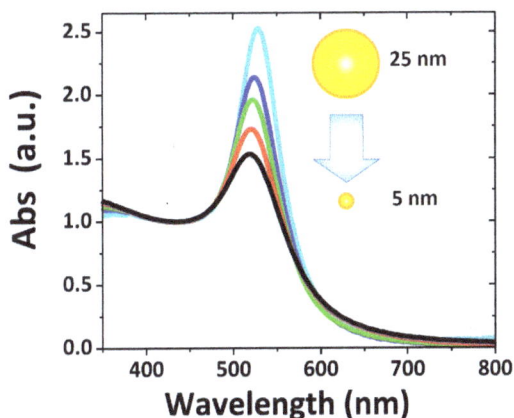

Figure 3. Example of intrinsic size effect observed in the AuNP SPR band for decreasing particle size from 25 nm to 5 nm. Reprinted with permission from (Amendola et al. 2017). Copyright® IOP Publishing.

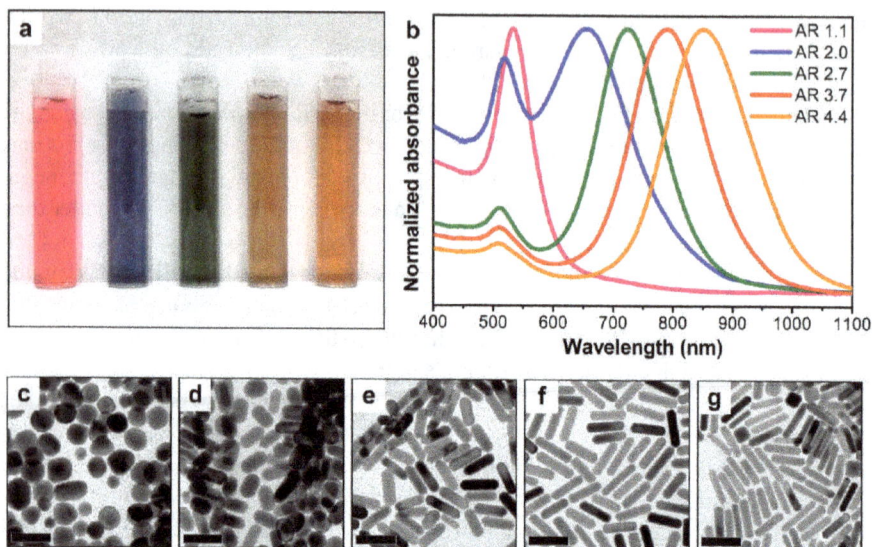

Figure 4. Influence of aspect ratio in the optical properties of AuNR. (a) Visible evidence of AuNR solutions with increasing aspect ratio left to right. (b) UV-vis absorbance spectra and (c–g) TEM images of AuNR with different aspect ratios (c) 1.1; (d) 2.0; (e) 2.7; (f) 3.7; and (g) 4.4. Scale bars = 50 nm. Reprinted with permission from (Abadeer et al. 2014). Copyright® American Chemical Society.

Another relevant factor is the dielectric property of the medium surrounding the nanoparticle. In a medium with a higher dielectric function, the polarizability of the MNP increases and the SPR band redshifts (Kinnan et al. 2009). The screening charges induced in the polarizable medium, in the vicinity of the nanoparticle surface, generate a "restoring force" to the oscillating electrons, which affect the frequency and intensity of the SPR. The larger the dielectric functions, the higher the medium polarizability. As a result of this, stronger restoring forces cause the SPR shift to lower energies (Kinnan et al. 2009). For example, when AuNP are dispersed in a medium with a larger refractive index, the SPR band would present the following characteristics: high intensity, redshift position, and broadening the maximum width to half-height (FWHM) (Valcárcel et al. 2014).

Another factor influencing the SPR is the interparticle distance. For instance, aggregates of spheres show a new SPR because the plasmons of individual particles begin to interact and hybridize with the plasmons of the close particles (Valcárcel et al. 2014). In the case of dimers, the hybridization of the SPR produces a redshift of the band (Figure 5). This effect decreases almost exponentially with the increase in the distance between particles, independently of the size, shape, metal type, and medium dielectric constant (Jain et al. 2007).

2.2.2 Luminescent Nanomaterials

Some NM behave as excellent luminophores, showing intense signals for fluorescence emission. This property makes them excellent candidates in lasers, optical sensors, and light-emitting diodes (LEDs) applications. There are two main mechanisms to

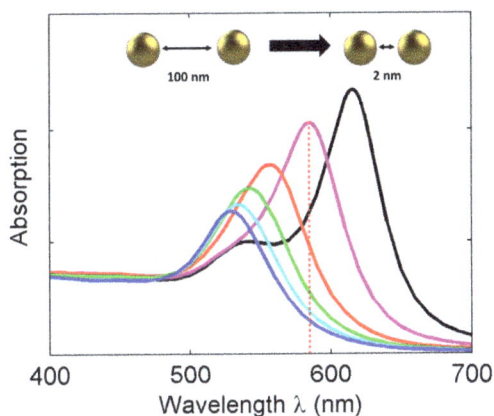

Figure 5. Influence of interparticle distance on the absorption spectra of AuNP dimers. Reprinted with permission from (Zhang et al. 2013). Copyright® the Optical Society.

produce light emission: photoinduction (*photoluminescence*, PL) in which a material absorbs a photon to reach an excited electronic state and then returns to a ground state by radiative emission of lower energy compared to the excitation one; and the second is electro-induction (*electroluminescence*, EL). In EL, the light emission is the result of electron-hole recombination produced after an electrical injection of the electron into the conduction band and the hole into the valence band (Zhang 2009). NM that show strong PL include passivated *quantum dots* (QDs) (Biju et al. 2008, Galian and Guardia 2009); carbon-dots (Sun and Lei 2017) and metal halide perovskites (Hoye et al. 2020, Schmidt et al. 2014, Shamsi et al. 2019).

In QDs, the fluorescence emission relates to their size and monodispersity. The smaller the size, the blueshift the emission. As the QDs are semiconductors, the light emission phenomenon is produced by recombination electron-hole (exciton) after photo-excitation of electrons in the valence band to the conduction band. The PL is enhanced by organic passivation of the QD or by modification to obtain core-shell structures QDs (Galian and Guardia 2009, Zhang 2009). Both modifications intend to decrease the presence of trap states and avoid emission quenching.

Carbon-dots are quasi-spherical NP comprising amorphous to crystalline regions in their structures containing sp^2/sp^3 carbon, oxygen/nitrogen based groups. Also, they can be modified to improve their luminescent properties. The size and surface functional groups modulate their fluorescence emission, which usually can be tuned by controlling the condensation and post-synthesis reactions, and by doping in other elements (Ghosal and Ghosh 2019, Sun and Lei 2017). The origin of their fluorescence is still unclear with the hypothesis that attributes the light emission to a similar basis as for QDs (quantum size effect, surface defects, emissive traps, and radiative recombination of the excitons), whilst more recent studies propose that carbon-dots have a distinctive hybrid combination of dye molecule-like and nano-semiconductors structures (Ghosal and Ghosh 2019).

Perovskites are formed by three primary ions, with an ABX_3 stoichiometry, where A is a monovalent cation (Cs or Rb), B is a divalent cation (Pb, Sn, Ge), and X is a halide ion (Cl, Br, I or a combination of them) (Shamsi et al. 2019). Among other properties, high absorption coefficients and PL quantum yields close to unity stand out. Besides, controlling their composition or size, the bandgap can be tuned over the entire visible electromagnetic region making them excellent candidates for photovoltaic and optoelectronics applications (Hoye et al. 2020).

2.3 Thermal Properties

Thermal conductivity, heat capacity, melting point, and glass transition temperature are some examples of thermal properties that are modified when the NM size decreases. Phase diagrams of NM depend on the particle size, shape, and the medium in which it is located. For example, the Gibbs-Thomson equation (Equation 1) describes the relationship between the melting point and the NM size,

$$T_{mp} = T_{mp}^* \left(1 - \frac{2\sigma_{sl}\kappa}{\rho_s L_f} \right)$$

(1)

where T_{mp}^* is the bulk melting point, σ_{sl} is the surface energy at the solid-liquid interface, L_f is the latent heat, ρ_s is the density, and κ corresponds to the curvature of the sphere (1/R, where R is the particle radius) (Font et al. 2015). Equation 1 predicts that the melting point depends inversely on R for spherical particles (Figure 6). At sizes smaller than 20 nm, the decrease in melting temperature is noticeable, and for nanoparticles less than 3 nm, it decreases to approximately 600 K.

Other materials, such as carbon nanotubes (CNTs), have superior thermal conductivity. It was observed experimentally that the thermal conductivity of CNTs can be as high as 3000 W/m/K, almost identical to the thermal conductivity of the

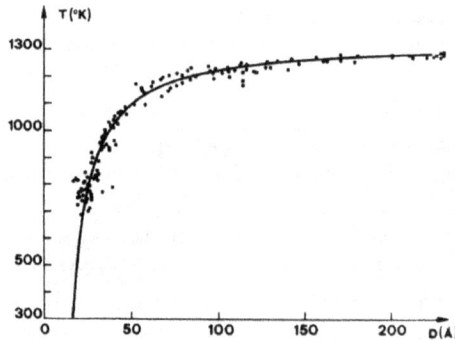

Figure 6. Melting point of AuNP as a function of their size. Reprinted with permission from (Buffat and Borel 1976). Copyright® American Physical Society.

diamond. Thermal conductivity of this magnitude allows CNTs to be used in circuits facilitating heat transport (Shi et al. 2015).

2.4 Magnetic Properties

Some materials show a response to an applied magnetic field and are classified as magnetic materials: ferromagnetic, paramagnetic, diamagnetic, antiferromagnetic, and ferrimagnetic.

Ferromagnetic materials, such as iron, cobalt, and nickel, have atoms with unpaired electrons; thus, they have a net magnetic moment. These materials are composed of magnetic domains (regions with atoms having parallel magnetic moments pointing in some direction). The application of an external magnetic field provokes the alignment of all the magnetic moments along its direction, forming a large net magnetic moment. After the magnetic field is removed, a residual magnetic moment persists (Issa et al. 2013).

Antiferromagnetic materials (e.g., MnO, CoO, NiO, and $CuCl_2$) are composed of two different elements, each having magnetic moments with equal magnitude, but pointing in the opposite direction, which results in zero net magnetic moments. In contrast, ferrimagnetic materials, such as magnetite (Fe_3O_4) and maghemite (γ-Fe_2O_3), have antiparallel magnetic moments that differ in magnitude; then, they do not cancel out, presenting a net spontaneous magnetic moment. Both materials behave similarly to the ferromagnetic ones in the presence of an external magnetic field (Issa et al. 2013).

When sample dimensions are significantly reduced, as in NP, the energetic stability achieved through the formation of domains decreases considerably. If no domains are formed, the system adopts a single-domain configuration, represented by a single 'superspin' (Thanh 2012). In ferromagnetic materials, the single-domain particles appear below a critical size of around 100 nm (Issa et al. 2013).

A *superparamagnetic* state appears when the magnetic moments fluctuate around the magnetization axes. This state takes place above the blocking temperature as the thermal energy is enough to overcome the anisotropic barrier causing magnetization to rapidly fluctuate from one orientation to another (Issa et al. 2013, Thanh 2012). Magnetic properties of NP are dominated by two main effects: finite-size (single-domain or multi-domain structures and quantum confinement of the electrons), and surface (Issa et al. 2013). The domain structures in magnetic NP control their magnetic behavior and size dependence. The particle shape, the crystal anisotropy strength, magnetic saturation, and the domain wall energy determine the critical size

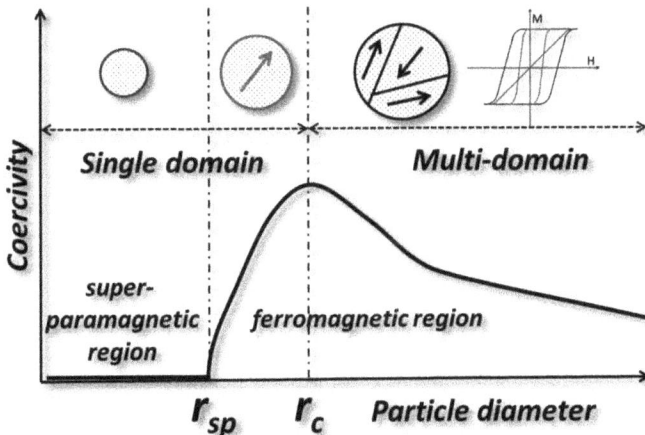

Figure 7. Size dependence for coercivity of magnetic nanoparticles. rc: critical radius, rsp: threshold radius for superparamagnetism. Reprinted with permission from (Kalubowilage et al. 2019). CC BY 4.0, https://doi.org/10.3390/app9142927, MDPI.

of single domains. The term coercivity defines the resistance of a magnetic material to magnetization variations, and it depends on the nanoparticle diameter (Figure 7) (Kalubowilage et al. 2019).

The surface effects, as a consequence of the large fraction of surface atoms, can lead to surface magnetization. For example, some bulk nonmagnetic materials such as CeO_2 and Al_2O_3 display in the nanoscale magnetic hysteresis loops at room temperature (Issa et al. 2013).

As mentioned above, this type of nanoparticles can be manipulated by an external magnetic field. This "remote action," combined with the intrinsic penetrability of biological tissues by a magnetic field, allows their application in biomedicine. For instance, they are used in applications involving the transport and/or immobilization of the nanoparticles, or the magnetic label of biological tissues. Also, its resonant response to a variable-time magnetic field can be used as a hyperthermia agent in tumors or as agents that promote chemotherapy and radiation therapy in cancer patients (Pankhurst et al. 2003).

2.5 Mechanical Properties

The superior mechanical properties in NM are a result of the high surface area. Thus, products containing nanomaterials have preferable strength, flexibility, and scalability than the bulk materials. In general terms, NM commonly double in strength and hardness to analogous of the same composition but larger size. For example, CNTs show high elastic modulus, high elastic strain and high rupture strain (Shi et al. 2015). Also, the addition of NMs such as nano-SiO_2 to a common material (e.g., concrete) improves its compressive, bending, and splitting tensile strengths. However, in some cases, excessive NM content can reduce the ordinary mechanical properties. This means that if enhanced mechanical properties are aimed, the type and ratio of nanoparticles added should be considered. Other relevant factors affecting these kinds of properties are the production process, grain size, and grain boundary structure (Wu et al. 2020).

Three categories summarize the most common mechanical behaviors: (i) in the case of polymeric NM, the size dependence is not uniform. For example, polystyrene NP (d = 200 nm) showed a lower compressive moduli compared with the bulk material whilst the elastic modulus in polypropylene NP increased with respect to the massive polymer; (ii) for MNP, one of the factors affecting their mechanical response is the dislocations inside the particles; and (iii) for nanowires and nanotubes, it was found that the elastic moduli decreased with the increasing radial diameter (Guo et al. 2014).

2.6 Chemical Properties

The reduction in the particle size also influences the NM performance in chemical reactions. The high surface area exposes more active sites to the environment which ultimately results in an increased reactivity. This is the reason why several NM are employed as catalysts in several chemical reactions (Thorat and Bauer 2020).

Catalysis is known as the process of introducing a chemical species to a reaction to increase its rate. This species, known as the *catalyst*, provides an alternative reaction pathway with a lower activation energy and it is not consumed during the reaction. Catalysis has enormous technological significance, by reducing reaction times and costs in energy production, the chemical industry, and environmental technology (Tao 2014). In recent years, the progress of nanoscience and nanotechnology led to the emergence of *nanocatalysis*, when it is mediated by nanoparticles, which is somehow between homogeneous (all the components are in a single phase) and heterogeneous (at least two phases are involved) catalysis (Zibareva et al. 2019).

To define nanocatalysis, two important conditions must be fulfilled. First, the valence electrons of the active part of a nanocatalyst are highly confined, leading to physical and chemical properties non-scalable from the bulk properties. As previously mentioned, this condition is true for particles in the nanometer length scale or smaller. Second, nanocatalysts are designed in a controlled manner, where the activity, specificity, and selectivity of the catalyst can be modified by modulating the particle size, by incorporating different atoms, changing the charge of the particle or modifying its magnetic properties (Rao et al. 2004).

The catalytic activity increases with decreasing the NM size. Noble metals in bulk are inactive and considered as poor catalysts. However, MNP act like very effective catalysts. For example, massive gold is considered a poor catalyst. Nevertheless, when its size is reduced to a few nm, these AuNP show enhanced catalytic activity in CO oxidation. Additionally, when AuNP are deposited on metal oxide like TiO_2, the combination of both shows improved catalytic activity than individual AuNP or nano-TiO_2. This is because CO gets adsorbed on the steps, edges, and corner sites of the AuNP, whilst it never gets adsorbed on the smooth surface of Au in bulk (Haruta and Daté 2001).

Furthermore, the catalytic selectivity depends on the size of the nanoparticle. Numerous investigations demonstrate that the catalytic activity can be enhanced from nanoparticle sizes, owing to their high surface area and surface roughness (An and Somorjai 2012). For instance, the size of PdNP supported on activated carbon significantly affects the catalyst performance on the selective hydrogenation of halonitrobenzenes, while the dehalogenation step does not. It was found that the selectivity to hydrogenation is over 99.90% when the PdNP were bigger than 25 nm (Lyu et al. 2014). Also, the size-dependent activity and selectivity of carbon dioxide photocatalytic reduction over PtNP has been investigated (Dong et al. 2018). They observed that decreasing the size of PtNP promotes the charge transfer efficiency, and thus enhances both the carbon dioxide photocatalytic reduction and hydrogen evolution reaction (HER) activity but leads to higher selectivity towards hydrogen over methane. Combining experimental results and theoretical calculations, in PtNP, the terrace sites are revealed as the active sites for methane generation; meanwhile, the low-coordinated sites are more favorable in the competing HER.

The shape of the NP is another parameter to consider in the catalyst's selectivity. For example, a particular selectivity for the conversion of *trans* olefins into their less thermodynamically stable *cis* isomers on close-packed Pt(111) surfaces was observed using tetrahedral nanoparticles, which only exposed (111) facets. Other nanoparticle shapes, by contrast, promote *trans*-olefin formation instead (Figure 8) (Lee and Zaera 2013). Nanoparticle shape selectivity is perhaps more critical in the promotion of the so-called mild reactions, which are carried out at low temperatures

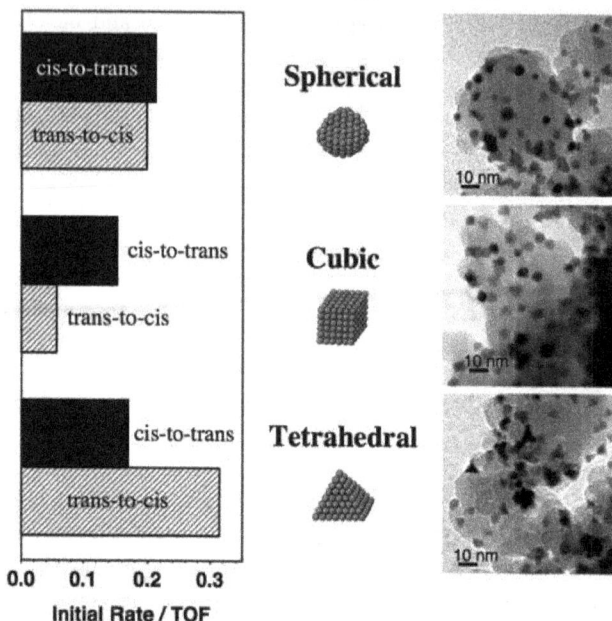

Figure 8. PtNP shape dependence for the selective isomerization of olefins. The data on the left corresponds to the initial rates as turnover frequencies (TOFs, in molecules/Pt atoms), for the conversion of *cis*-2-butene to the *trans*-isomer (solid blue) and for the opposite reaction (hatched red). In all cases, the PtNP sizes are around 5 nm in diameter. In addition, the TEM images are shown on the right. Reprinted with permission from (Lee et al. 2011). Copyright® Royal Society of Chemistry.

and atmospheric pressures. More demanding reactions such as ammonia synthesis, Fischer–Tropsch, water–gas shift, and most oxidations are known to be structure sensitive and could, in principle, be excellent candidates for this type of selectivity control based on the nanoparticles' shape (Lee and Zaera 2013). However, the stringent conditions required for those catalytic conversions may lead to the loss of the initial particle shapes, and with that to a deterioration in catalytic performance.

Moreover, as certain NP are thermodynamically unstable, capping agents are always added to stabilize them by steric repulsion, electrostatic repulsion or a combination of both (Liu and Jiang 2015). The presence of the ligands could modify various functional groups and charges on the NP surface, and consequently influence their reactivity.

The *electrostatic stabilization* comprises the formation of a double-diffusion layer when NP are dispersed in a solvent. When two particles approach each other, the double layers of the particles overlap, which generates a repulsive force and avoid agglomeration (Thorat and Bauer 2020). The repulsive force decreases with increasing separation between the particles. The possibility and efficiency of electrostatic stabilization can be evaluated by measuring the zeta potential as a function of pH (Hang et al. 2009).

The adsorption of large molecules on the nanoparticle surface, including surfactants or polymers, to prevent aggregation is called **steric stabilization**. The osmotic repulsive force augments due to increasing the concentration of adsorbed molecules, and then the stability of nanostructured particles enhances. Hence, the interaction potential can be extended to point out the repulsive forces that support steric stabilization. The stability due to steric factors depends on temperature, polymer concentration, solubility, and average chain length. As it is not a long-range interaction, it does not depend on the particle size (Mohapatra et al. 2018).

The ligand shell surrounding the particles is the region where most of the chemical and physical events of NP with other molecules or more complex chemical assemblies take place. A monolayer of ligands around these particles inherits the curvature of the core; as a consequence, there is enough room between different chains comprising the monolayer to embed other small molecules. The monolayers are expected to exhibit a host-guest chemistry on their own, in addition to a more selective host-guest chemistry imparted by a specific design of the monomers comprising the monolayer (Valcárcel et al. 2014). For instance, AuNP may be stabilized by interaction with compounds such as thiols that form a compact, self-assembled monolayer on their surface (Valcárcel et al. 2014). One of the most important chemical properties of the Au-S interface of alkanethiolate-protected AuNP is the possibility to sustain exchange reactions by replacing a ligand with another one without affecting the structural integrity of the particles. In fact, many reports on the preparation of particles featuring mixed monolayers exploit this synthetic route. For instance, a new ligand exchange method for cetyltrimethylammonium bromide (CTAB) or chloride (CTAC) stabilized AuNP was developed (Dewi et al. 2014). The CTAB/C surfactants could be easily removed via dichloromethane solvent extraction whilst maintaining colloidal stability, and subsequent ligand replacement with 3-mercaptopropionic acid (3-MPA) resulted in a complete exchange of CTAB/C with 3-MPA. The resulting 3-MPA-capped AuNP feature very small hydrodynamic diameters and exhibit excellent colloidal stability (Dewi et al. 2014). Moreover, ligand exchange reactions could help modulate the structure and reactivity of certain complexes to achieve more efficient catalysts. For example, colloidal ligand-exchange methods were used to adsorb a single layer of tetrathiometallate complex (MoS_4^{2-}, WS_4^{2-}) onto colloidal AuNP, and characterize the influence of the Au support on the electronic and geometric properties of the surface MSx monolayer (Yadav et al. 2020). Utilizing this kind of method ensures that every MoS_x active site interacts directly with the Au surface, and as a consequence, the sites experience strong electronic and geometric perturbations. The Au surface plays a crucial role in templating a densely packed, oligomerized MoS_x structure whose molecular orbitals are strongly hybridized with those of the Au surface. Both factors play a role in weakening hydrogen atom binding energies for more optimal catalysis and stabilizing the HER catalytic turnover (Yadav et al. 2020).

Conclusions

The multiple applications of NMs are a direct consequence of their unique physical and chemical properties. A good knowledge on how these properties can be controlled is essential to design new NM with custom-made characteristics. This chapter summarized the most relevant properties to consider when working with NM. Most of them are dominated by size, but shape and surface ligands also play a relevant role. Far from being completely understood, there is still room for basic research in understanding different photophysical phenomena like the fluorescence origin of carbon-dots, the fundamental properties of perovskites or the mechanical properties of different NM. We expect this chapter to be useful as an introduction to the most relevant physical and chemical properties for those who are interested on beginning to work with NM or would like to refresh some basic concepts.

Acknowledgements

Thanks to all the scientific community whose relevant work is referenced in this chapter. N.L.P. is a research member of the CONICET, Argentina. M.A.M.T. is a grateful recipient of a postdoctoral fellow from CONICET. This work has been possible thanks to financial support by the CONICET, PIP 2017-2019 Code 11220170100505CO and the Secyt-UNC 2018-2021, Res. 411/18.

References

Abadeer, N.S., Brennan, M.R., Wilson, W.L. and Murphy, C.J. (2014). Distance and plasmon wavelength dependent fluorescence of molecules bound to silica-coated gold nanorods. *ACS Nano*, 8(8): 8392–8406. https://doi.org/10.1021/nn502887j.

Amendola, V., Pilot, R., Frasconi, M., Maragò, O.M. and Iatì, M.A. (2017). Surface plasmon resonance in gold nanoparticles: A review. *Journal of Physics: Condensed Matter*, 29(20): 203002. https://doi.org/10.1088/1361-648X/aa60f3.

An, K. and Somorjai, G.A. (2012). Size and Shape control of metal nanoparticles for reaction selectivity in catalysis. *ChemCatChem*, 4(10): 1512–1524. https://doi.org/10.1002/cctc.201200229.

Auffan, M., Rose, J., Bottero, J., Lowry, G., Jolivet, J. and Wiesner, M. (2009). Towards a definition of inorganic nanoparticles from an environmental, health and safety perspective. *Nature Nanotechnology*, 4(10): 634–641. https://doi.org/10.1038/nnano.2009.242.

Biju, V., Itoh, T., Anas, A., Sujith, A. and Ishikawa, M. (2008). Semiconductor quantum dots and metal nanoparticles: Syntheses, optical properties, and biological applications. *Analytical and Bioanalytical Chemistry*, 391(7): 2469–2495. https://doi.org/10.1007/s00216-008-2185-7.

Buffat, Ph. and Borel, J.-P. (1976). Size effect on the melting temperature of gold particles. *Physical Review A*, 13(6): 2287–2298. https://doi.org/10.1103/PhysRevA.13.2287.

Buzea, C., Pacheco, I. and Robbie, K. (2007). Nanomaterials and nanoparticles: Sources and toxicity. *Biointerphases*, 2(4): MR17–MR71. https://doi.org/10.1116/1.2815690.

Coronado, E., Encina, E. and Stefani, F. (2011). Optical properties of metallic nanoparticles: Manipulating light, heat and forces at the nanoscale. *Nanoscale*, 3(10): 4042. https://doi.org/10.1039/c1nr10788g.

Dewi, M.R., Laufersky, G. and Nann, T. (2014). A highly efficient ligand exchange reaction on gold nanoparticles: Preserving their size, shape and colloidal stability. *RSC Advances*, 4(64): 34217–34220. https://doi.org/10.1039/c4ra05035e.

Dong, C., Lian, C., Hu, S., Deng, Z., Gong, J., Li, M., Liu, H., Xing, M. and Zhang, J. (2018). Size-dependent activity and selectivity of carbon dioxide photocatalytic reduction over platinum nanoparticles. *Nature Communications*, 9: 1252. https://doi.org/10.1038/s41467-018-03666-2.

Font, F., Myers, T.G. and Mitchell, S.L. (2015). A mathematical model for nanoparticle melting with density change. *Microfluid Nanofluid*, 18: 233–243. https://doi.org/DOI 10.1007/s10404-014-1423-x.

Galian, R. and Guardia, M. (2009). The use of quantum dots in organic chemistry. *TrAC Trends in Analytical Chemistry*, 28(3): 279–291. https://doi.org/10.1016/j.trac.2008.12.001.

Ghosal, K. and Ghosh, A. (2019). Carbon dots: The next generation platform for biomedical applications. *Materials Science and Engineering: C*, 96: 887–903. https://doi.org/10.1016/j.msec.2018.11.060.

Ghosh, S. and Pal, T. (2007). Interparticle coupling effect on the surface plasmon resonance of gold nanoparticles: From theory to applications. *Chemical Reviews*, 107(11): 4797–4862. https://doi.org/10.1021/cr0680282.

Guo, D., Xie, G. and Luo, J. (2014). Mechanical properties of nanoparticles: Basics and applications. *Journal of Physics D: Applied Physics*, 47(1): 013001. https://doi.org/10.1088/0022-3727/47/1/013001.

Hang, J., Shi, L., Feng, X. and Xiao, L. (2009). Electrostatic and electrosteric stabilization of aqueous suspensions of barite nanoparticles. *Powder Technology*, 192(2): 166–170. https://doi.org/10.1016/j.powtec.2008.12.010.

Haruta, M. and Daté, M. (2001). Advances in the catalysis of Au nanoparticles. *Applied Catalysis A: General*, 222: 427–437. https://doi.org/10.1016/S0926-860X(01)00847-X.

Hoye, R.L.Z., Fakharuddin, A., Congreve, D.N., Wang, J. and Schmidt-Mende, L. (2020). Light emission from perovskite materials. *APL Materials*, 8(7): 070401. https://doi.org/10.1063/5.0019554.

Issa, B., Obaidat, I., Albiss, B. and Haik, Y. (2013). Magnetic nanoparticles: Surface effects and properties related to biomedicine applications. *International Journal of Molecular Sciences*, 14(11): 21266–21305. https://doi.org/10.3390/ijms141121266.

Jain, P., Huang, W. and El-Sayed, M. (2007). On the universal scaling behavior of the distance decay of plasmon coupling in metal nanoparticle Pairs: A plasmon ruler equation. *Nano Lett.*, 7(7): 2080–2088. https://doi.org/10.1021/nl071008a.

Kalubowilage, M., Janik, K. and Bossmann, S.H. (2019). Magnetic nanomaterials for magnetically-aided drug delivery and hyperthermia. *Applied Sciences*, 9(14): 2927. https://doi.org/10.3390/app9142927.

Kinnan, M.K., Kachan, S., Simmons, C.K. and Chumanov, G. (2009). Plasmon coupling in two-dimensional arrays of silver nanoparticles: I. Effect of the dielectric medium. *The Journal of Physical Chemistry C*, 113(17): 7079–7084. https://doi.org/10.1021/jp900090a.

Lakowicz, J. (2005). Radiative decay engineering 5: Metal-enhanced fluorescence and plasmon emission. *Analytical Biochemistry*, 337(2): 171–194. https://doi.org/10.1016/j.ab.2004.11.026.

Lee, I., Albiter, M.A., Zhang, Q., Ge, J., Yin, Y. and Zaera, F. (2011). New nanostructured heterogeneous catalysts with increased selectivity and stability. *Physical Chemistry Chemical Physics*, 13: 2449–2456. https://doi.org/10.1039/c0cp01688h.

Lee, I. and Zaera, F. (2013). Nanoparticle shape selectivity in catalysis: Butene isomerization and hydrogenation on platinum. *Topics in Catalysis*, 56: 1284–1298. https://doi.org/10.1007/s11244-013-0155-6.

Lee, S.Y., Tsalu, P.V., Kim, G.W., Seo, M.J., Hong, J.W. and Ha, J.W. (2019). Tuning chemical interface damping: interfacial electronic effects of adsorbate molecules and sharp tips of single gold bipyramids. *Nano Lett.*, 19: 2568–2574. https://doi.org/10.1021/acs.nanolett.9b00338.

Link, S. and El-Sayed, M. (1999). Spectral properties and relaxation dynamics of surface plasmon electronic oscillations in gold and silver nanodots and nanorods. *The Journal of Physical Chemistry B*, 103(40): 8410–8426. https://doi.org/10.1021/jp9917648.

Liu, J. and Jiang, G. (eds.). (2015). *Silver Nanoparticles in the Environment* (1st ed.). Springer. https://doi.org/10.1007/978-3-662-46070-2.

Lyu, J., Wang, J., Lu, C., Ma, L., Zhang, Q., He, X. and Li, X. (2014). Size-dependent halogenated nitrobenzene hydrogenation selectivity of Pd nanoparticles. *J. Phys. Chem. C*, 118: 2594–2601. https://doi.org/10.1021/jp411442f.

Mohapatra, S., Ranjan, S., Dasgupta, N., Mishra, R. and Thomas, S. (eds.). (2018). *Characterization and Biology of Nanomaterials for Drug Delivery Nanoscience and Nanotechnology in Drug Delivery* (1st ed.). Elsevier.

Pankhurst, Q.A., Connolly, J., Jones, S.K. and Dobson, J. (2003). Applications of magnetic nanoparticles in biomedicine. *J. Physics D: Applied Physics*, 36: R167–R181.

Peiris, S., McMurtrie, J. and Zhu, H.-Y. (2016). Metal nanoparticle photocatalysts: Emerging processes for green organic synthesis. *Catalysis Science & Technology*, 6(2): 320–338. https://doi.org/10.1039/C5CY02048D.

Rao, C.N.R., Müller, A. and Cheetham, A.K. (eds.). (2004). *The Chemistry of Nanomaterials: Synthesis, Properties and Applications* (1st ed.). Wiley-VCH.

Schmidt, L.C., Pertegás, A., González-Carrero, S., Malinkiewicz, O., Agouram, S., Mínguez Espallargas, G., Bolink, H.J., Galian, R.E. and Pérez-Prieto, J. (2014). Nontemplate Synthesis of CH$_3$NH$_3$PbBr$_3$ Perovskite Nanoparticles. *Journal of the American Chemical Society*, 136(3): 850–853. https://doi. org/10.1021/ja4109209.

Shamsi, J., Urban, A.S., Imran, M., De Trizio, L. and Manna, L. (2019). Metal halide perovskite nanocrystals: Synthesis, post-synthesis modifications, and their optical properties. *Chemical Reviews*, 119(5): 3296–3348. https://doi.org/10.1021/acs.chemrev.8b00644.

Shi, D., Guo, Z. and Bedford, N. (2015). *Nanomaterials and Devices* (1st ed.). Elsevier Ltd.

Sun, X. and Lei, Y. (2017). Fluorescent carbon dots and their sensing applications. *TrAC Trends in Analytical Chemistry*, 89: 163–180. https://doi.org/10.1016/j.trac.2017.02.001.

Tao, F. (ed.). (2014). *Metal Nanoparticles for Catalysis: Advances and Applications*. Royal Society of Chemistry. https://doi.org/10.1039/9781782621034

Thanh, N.T.K. (ed.). (2012). *Magnetic Nanoparticles. From Fabrication to Clinical Applications*. CRC Press.

Thorat, N.D. and Bauer, J. (eds.). (2020). *Nanomedicines for Breast Cancer Theranostics*. Elsevier. https://www.sciencedirect.com/science/book/9780128200162.

Valcárcel, M., López-Lorente, Á.I., Barceló, D. and Wilson, C.L. (eds.). (2014). *Gold Nanoparticles in Analytical Chemistry*. Elsevier.

Wu, Q., Miao, W., Zhang, Y., Gao, H. and Hui, D. (2020). Mechanical properties of nanomaterials: A review. *Nanotechnology Reviews*, 9(1): 259–273. https://doi.org/10.1515/ntrev-2020-0021.

Yadav, V., Lowe, J.S., Shumski, A.J., Liu, E.Z., Greeley, J. and Li, C.W. (2020). Modulating the structure and hydrogen evolution reactivity of metal chalcogenide complexes through ligand exchange onto colloidal Au nanoparticles. *ACS Catal.*, 10: 13305–13313. https://doi.org/10.1021/acscatal.0c02895.

Zhang, J.Z. (2009). *Optical Properties and Spectroscopy of Nanomaterials*. World Scientific.

Zhang, W., Li, Q. and Qiu, M. (2013). A plasmon ruler based on nanoscale photothermal effect. *Optics Express*, 21(1): 172. https://doi.org/10.1364/OE.21.000172.

Zibareva, I.V., Ilina, L.Y. and Vedyagin, A.A. (2019). Catalysis by nanoparticles: The main features and trends. *Reaction Kinetics, Mechanisms and Catalysis*, 127(1): 19–24. https://doi.org/10.1007/ s11144-019-01552-6.

Chapter 3

Characterization of Nanomaterials

Nora Catalina Restrepo-Zapata

1. Introduction

Currently, the instrumental techniques to characterize nanomaterials are innumerable, but the analysis always depends on the experience to determine the most suitable approach for the objective of a research or industry project. This chapter will be divided into categories of properties to be characterized as follows:

- Physical and chemical properties
- Thermal properties
- Rheological properties
- Mechanical properties
- Magnetic properties
- Electrical properties
- Optical properties

The different tasks before sample preparation and subsequent data analysis to the measurement must be considered for successful nanomaterial characterization. Depending on the technique, some may be simple and quick, but others can take days to complete. Also, they rely on the correct development of the characterization and subsequent analysis. Since there is a wide variety of types of nanomaterials, it will be difficult to generalize; however, the best effort is made to give a common knowledge for sample preparation and characterization. The authors recommend following the suggested literature to go deeper into a specific topic.

Universidad Autónoma de Chile, Facultad de Ingeniería, Ingeniería Civil Química. Avenida Pedro de Valdivia 425, Providencia, Santiago, Chile.
Email: nora.restrepo@uautonoma.cl

2. Sample Preparation

This activity includes size reduction, dissolution, substance isolation, and interference removal. On certain occasions, a comparison between samples will be added to facilitate quantifying and estimating the nanomaterial under analysis. Usually, extraction, dissolution, and even pressing techniques are required, depending on the type of characterization. Table 1 summarizes the main techniques, objectives, and uses.

Table 1. Nanoparticle sample preparation techniques.

Technique	Objective	Description
Dispersion and stabilization (ASTM 2021)	Preserves the colloidal stability of the nanomaterial during the measurement	A suitable diluent called "mother liquor" is used in which the nanomaterial exists in a stable form and is obtained from the centrifugation of the suspension or by performing, if possible, a dispersing liquid of the same ionic nature. The nanomaterial is dispersed to the appropriate concentration for measurement.
Deposition (ASTM 2017)	Provides a consistent baseline for flat or height measurements	High-quality substrates such as mica, atomically flat gold (111) deposited on mica, or monocrystalline silicon are used to minimize the effect of surface roughness on measurement.
Solution deposition (Baer 2020)	Provides a film that completely covers a substrate to eliminate any signs of interference	Depending on the specific technique, a solution is added through drop coating, spin coating, and solution dipping. The solution is invisible to the technique or does not generate noise in the result.
Washing and separation (Baer 2020)	Separates and cleans supplied nanomaterials in a solution	Dialysis, diafiltration, centrifugation, and filtering are used to obtain nanomaterials free of impurities or remove excess solution. Once separated, they can be dried or agglomerated according to the technique used.
Thin-film (Zhang 2020)	Forms a characteristic, pure, and well-defined sample for physical and chemical techniques, preferably	A cut is made using a cooled microtome, either cryogenically or refrigerant, that does not react with the sample, through a specific thickness or area.
Digestion (Anton Paar 2022)	Reduces and dissolves nanomaterials or nanomaterial compounds to facilitate their analysis. It is preferred for preparing spectrometric samples	Concentrated acids or mixtures destroy or dissolve organic or inorganic compounds, leaving only the compound to be investigated. Also, it can be used to convert a solid material into a dispersion.

The Material Measurement Laboratory of the National Institute of Standards and Technology has on its website (https://www.nist.gov/mml/nano-measurements-complete-list-protocols) (NIST 2022) a detailed description of the different sample preparation protocols according to their nature, application, and chemical composition, which may be of interest if you want to go deeper into this topic.

3. Physical and Chemical Properties

Typical classifications of substances are based on their physical and chemical properties. The main parameters of nanomaterials are shape, size, size distribution, physical structure, chemical composition, crystallographic structure, surface description, association, interface, and topology, which are explained in depth in the ASTM E3144 (ASTM 2019) and ASTM E3206 (ASTM 2019) standards.

3.1 Form

The characterization of the geometric shape of a nanomaterial is crucial since it affects most of its properties and reactivity, which must be considered when designing any nanomaterial. Quantitatively, it can be defined by the ratio of appearance, flatness, and sphericity (ASTM 2019). The form also affects its interaction with the surroundings or in the environment in which it is suspended. In general, nanomaterials are spherical or cylindrical; however, this can deviate into hollow shapes, stars, and even amorphous depending on the needs.

Transmission Electron Microscopy (TEM) is a technique that uses an electron beam to generate the image of a nanomaterial providing better resolution than optical incident light techniques. In addition to the shape of the particle, the size and distribution can also be determined using data acquisition software. This technique was used to systematically understand nanoparticle shape-dependent effects to predict properties and applications systematically (Lohse et al. 2014). Changes in the concentration of surfactant, silver salts, silver nitrate, and ascorbic acid during synthesis in the formation of nanoparticles of silver generate anisotropic nanomaterials, affecting the shape, too. It was identified that the halide ions affect the redox potentials, which affects the silver underpotential deposition in the nanocrystals (nanoseeds).

Another technique is Scanning Electron Microscopy (SEM). This microscope image is also formed with an electron beam from a tungsten filament, which interacts directly with the atoms. It generates a signal of low-energy secondary electrons that penetrate the sample at different depths and present the surface texture of the nanomaterial. The difference with TEM lies in the type of electrons used to form the image (for TEM, they pass through the sample). There are countless applications for SEM and green chemistry, regardless of whether it is based on biological systems, which is no exception. In the study by Fakhari, zinc oxide nanoparticles were made from *Laurus nobilis* L. and two salts, zinc acetate and nitrate (Fakhari et al. 2019). It is observed that for both zinc salts, spherical nanoparticles are generated. With zinc acetate, the growth is slower, so the size is smaller, generating aggregates similar to bullets. In turn, nitrate generates larger nanoparticles and forms flower-shaped bundles.

3.2 Size

It is explicitly known that nanomaterials' volume is in the range between 1 nm^3 to 10^6 nm^3 (definition of a nanometer), but these dimensions in terms of surface area will affect any characteristics on a larger scale (ASTM 2019). Photon Correlation Spectroscopy (PCS), also called Dynamic Light Scattering (DLS) or

Quasi-Elastic Light Scattering (QELS), is used to measure particle size within a liquid medium. This technique uses a laser projected to the suspension for a certain time. Due to Brownian motion, light is scattered, and the amount of scattering and time is recorded by a detector (Anton Paar 2022). The data acquisition system performs a correlation function with respect to time, obtaining the polydispersity index (PDI) and the particle size distribution. These procedures are standardized by ISO 22412 (ISO 2017) and ASTM E2490 (ASTM 2021). Depending on the angle of the detector, it can be determined if the nanomaterial is in aggregate form (15°), has some large particle sizes (90°), and backscattering (175°), or has wide polydispersity.

Baradi et al. (Barabadi et al. 2021) used this technique to determine the size of nanoparticles obtained from *Zakaria multiflora* extract at different pH values and thus compare them with commercial silver nanoparticles. Their antibacterial and biofilm inhibitory activity was also observed. It showed a decrease in hydrodynamic particle size with increased pH and obtained polydispersity indices below 0.7, indicating the formation of a monodisperse solution. The pH 9 nanoparticles were chosen from these four analyses since they provide advantages for easy penetration of the cell membrane and allow functionalizing the membrane to be used as a targeted drug delivery system. On the other hand, Amaro-Gahete and colleagues used DLS to calculate the hydrodynamic radius of graphene nanosheets (Amaro-Gahete et al. 2019). Although a nanosheet is not spherical, the information obtained shows its apparent size in the aqueous dispersion. In addition, it provides the polydispersity index, which for a nanosheet is obtained from the relationship between the width and the square of its height. In this case, as a new method of obtaining was analyzed, the sample has a broad particle size distribution as expected from conventional methods.

Nanoparticle Tracking Analysis (NTA) is a popular method for the direct and real-time visualization of suspended nanoparticles. A laser illuminates the particles, and the scattered light is recorded and magnified by a digital optical system. The software analyzes the video, obtaining each particle's movement and the size distribution. Unlike PCS, in NTA, the size distribution is a direct measurement (ASTM 2018). Luo's research team used this technique to determine the potential environmental impact of nanoparticles on natural sedimentary water systems (Luo et al. 2018). Different samples from the Derwent River in the UK were analyzed and stored for three months at 4°C. Then, they were mixed with citrate and mercaptoundecanoic acid coating gold nanoparticles. The study showed the presence of agglomerates for the two types of mixed nanoparticles and differed from the particles due to the characteristic sediments of the river. For real environmental studies, it would be a simple and fast technique to determine contamination and take appropriate actions to avoid or at least control it.

Atomic Force Microscopy (AFM) was one of the first techniques used to quantitatively determine the size of nanomaterials in dry form. This technique's advantage is that it provides a three-dimensional surface profile with accuracy and precision. AFM uses a sharp probe in a cantilever attached to a piezoelectric displacement actuator. The probe tip interacts with the surface experiencing an attractive or repulsive force creating a bending momentum. This deflection is detected by a laser and will correspond to the height. The piezoelectric displacement is recorded as a function of the position (ASTM 2017). This technique was used by Diculescu and coleagues to analyze the structure obtained in electrodes modified with nanoparticles to change the surface area and decrease the overpotential necessary for the redox reaction (Diculescu et al. 2007). For this, a solution of $PdCl_2$ with palladium nanowires electrochemically deposited on a surface of highly oriented pyrolytic graphite (HOPG) was used. Using AFM in Magnetic AC (AMC) mode, branches of palladium nanowires were observed, and each nanowire was made up of tiny nanoparticles. This confirms that the deposition was successful, and a honeycomb structure was created over all the surface defects of the HOPG.

Small Angle X-ray Scattering (SAXS) is another analytical technique that provides information about the structure and morphology of nanomaterials. It can be performed statically or dynamically. For this, the size and shape can be measured. In addition to the concentration in dispersion, it can characterize the particles' growth mechanism and assembly and determine the nanomaterial's alignment during processing. A high-resolution detector registers how an X-ray is dispersed in a sample based on the scattering angle. As its name implies, measurements are made at small angles, typically in the range of 0.1° and 5°. Since a scattering curve is obtained with the technique, quantitative data is obtained through data analysis. Rattanawongwiboon and collaborators used this technique to study the particle size and size distribution of gold nanoparticles obtained from a green synthesis of $HAuCl_4$ and a water-soluble chitosan solution. Increasing the concentration of $HAuCl_4$ produces larger particle sizes (Rattanawongwiboon et al. 2022). This was also complemented with TEM since the SAXS detected only the gold core but not the organic coating generated by the chitosan. It is clear that experience is necessary to determine which techniques can complement the results, especially when designing new nanomaterials.

Differential Centrifugal Sedimentation (DCS) is a new technique used to determine particle size and its distribution in nanomaterials between 2 nm and 80 microns (depending on the density of the nanomaterial). It complements the DLS technique that focuses on larger nanomaterial sizes. The sedimentator is typically a hollow disk in which a short wavelength monochromatic beam of light hits. The disk rotates up to 24,000 rpm, and the detector identifies different bands formed in the sample. Each band corresponds to each particle size, and the information becomes a curve that relates the relative weight and particle diameter. According to the change of variables during the test, the nanoparticle coating thickness, pore size distribution, and morphology disruptions can also be calculated (Analytik 2022). In the quality control of nanoparticles used in healthcare, DCS has been used to detect altered biomolecular crowns in inorganic nanoparticles of gold, CdSn/ZnS, silver, and iron platinum coated with poly-(isobutylene-alt-maleic anhydride)-graft-dodecyl (PMA). These types of modifications affect the physicochemical properties of nanoparticles. The presence of the PMA coating slightly moves the size distribution to smaller values due to biomolecular crown generation. However, the change in density is not significant, so the effect on performance would be minimal (Perez-Potti et al. 2021).

3.3 Physical Structure

The physical structure of a nanomaterial depends on its complexity in terms of homogeneities, holes, protrusions, appendages, or shell-like structures (ASTM 2019). In addition to techniques such as TEM, SEM, AFM, and SAXS that give information on size and shape, others focus solely on the physical structure.

Electron Backscatter Diffraction (EBSD) is a widely used technique for performing quantitative microstructural analysis, generally coupled to a SEM. This technique uses the incident electrons backscattered outside the surface generating a diffractogram with the atomic planes. The detector is a phosphor screen coupled to a camera that detects fluorescence, creating the image. The image or pattern obtained consists of bright bands that intersect themselves. Each band is produced by a series of crystalline planes within the sample. This limits its use to crystalline nanomaterials; however, during the design of nanomaterials, it will be able to give information about their atomic physical organization (pattern → crystalline; without a pattern → amorphous). The width of each band is related to the *d* spacing of the planes, and the angle between them corresponds to the angle between planes. To analyze nanomaterials with this technique, they must be on a flat surface, and if they are in suspension, they must be dried and mounted on clean substrates that do not generate noise during the analysis. Polished monocrystals are recommended in a known direction of a plane (Zainab 2015).

An interesting application of this technique was the determination of the microstructure of a nickel matrix nanocomposite produced by powder metallurgy and reinforced with carbon nanotubes. Here SEM polarized microscopy, EBSD, and metallography techniques were mixed to create a surface map of the nickel nanoparticle. It is observed that the nanoparticle is formed by several grains without preferential orientation or texture, which results from plastic deformation that occurs during the production of nickel powders (Carneiro et al. 2019).

3.4 Chemical Composition

Chemical characterization is the natural way to describe nanomaterials. The chemical composition can be expressed in terms of the molecules or atoms in a material, as a percentage of the material, or even using chemical bonds (ASTM 2019).

X-Ray Diffraction (XRD) is used to characterize nanomaterials of any size. Changes in diffraction peaks determine the crystal structure and how its cell parameters change regarding its size and shape. An X-ray hits electrons of the atoms of the material generating a regular arrangement of spherical waves. Using data acquisition software and Bragg's law, the scattering ray obtains the spacing between the crystalline planes and their respective angle (TWI 2022). This technique is widely used in materials science coupled with SEM or TEM. For example, XRD was used to analyze the photocatalytic response of bionanocomposites formed by TiO_2 nanotubes "decorated" with electrochemically deposited copper nanoparticles. Diffraction peaks were found at 36.2, 42.2, and 61.4° corresponding to the planes <111>, <200>, and <220> of Cu_2O, which confirms that the methodology used to synthesize the decorated nanotubes improves the photoactivity of the semiconductor by reducing its band gap energy (Sarto et al. 2019).

X-Ray Absorption Spectroscopy (XAS) is a chemical analysis technique that involves measuring the transmitted X-rays or their respective fluorescence as a function of increasing the energy of the X-rays; in simple terms, the technique measures the absorbed energy required to expel an electron from the electron shell of an element of interest. Depending on the absorption region, the name of the technique changes: for energies close to the absorption edge, it is called X-Ray Absorption Near-Edge Spectroscopy (XANES), while for the regions greater than the absorption edge, it is called X-Ray Absorption Fine Structure (EXAFS) (SIGRAY 2022). Energy calibration is made in terms of a specific atom depending on the characterization needs. The spectrum is collected and normalized using data acquisition software. This technique was used to observe nanoplatforms for photocatalysis using plasmonic Ag@TiO$_2$ core-shell isolated nanoparticles graphitized superficially with two complexes: Cu(I)/bipyridine and Cu(II)/bipyridine. The first one was mixed with boron nitride in an argon atmosphere to prevent oxidation in Cu(I). A copper calibration was performed using the XAS transmission mode, and the spectrum was measured in energy values close to the absorption edge (XANES) in the fluorescence mode. The spectrum shows differences between the two nanomaterials in shape and features of the absorption edge since, for the sample with Cu(I), a peak appears at the beginning of the ascent slope, which gives valuable information about the oxidative mechanism during catalysis and the plasmonic effects related to ligand's coordination to the metal ion (Queffélec et al. 2020).

One of the most used techniques to determine the chemical composition of nanomaterials is X-Ray Photoelectron Spectroscopy (XPS), also called Electron Spectroscopy for Chemical Analysis (ESCA), since it identifies the nature and surface consistency. For example, it can also determine the structure and thickness of nanoparticle coating (Baer 2020). Here, the sample's constituent elements, composition, and chemical bond are analyzed using incident X-rays on the surface and measuring the energy of the emitted photoelectrons. Hence, one of the main disadvantages of this technique is the meticulous preparation of the sample since any impurity would generate inappropriate information in the spectrum.

In the electrochemical synthesis of polypyrrole nanorods and nanoparticles, suitable electrolytic solutions are used to continuously polymerize the monomer and generate the nanostructures. However, this form of synthesis favors the superficial attack of certain substances used during synthesis, changing the physicochemical properties of the polymer material. For this purpose, nanorods and nanoparticles were exposed to ammonia to observe changes in chemical states and electrons through XPS. The N 1s core level peak was evaluated, evidencing a transfer mechanism by increasing the polarons (nitrogen ions charged positively (-N-H$^+$)). At the same time, the bipolarons decrease (=N-H$^+$). This means that impurities are being incorporated during processing, which will adversely affect the material (Šetka et al. 2019).

Low-Energy Ions Scattering (LEIS), commonly called Ion Scattering Spectroscopy (ISS), is a surface analytical technique that has the sole purpose of determining the elemental composition of the upper atomic layer of a sample. Unlike the previous X-rays techniques, this one uses an ideal gas ion beam scattered on the sample's surface. The scattered ions' kinetic energy is measured using conservation and momentum laws, obtaining peaks in a spectrum. The information corresponds to the elastic scattering of the ions of surface atoms. New platinum-gold bimetallic systems for fuel cells were characterized by LEIS to selectively analyze the atomic composition of the outer layer of a catalyst with two different compositions. The type of atoms found in it largely determines its activity and selectivity. Their spectra were compared with gold and platinum obtained under the same conditions. It was observed that catalyst A has a higher gold signal; however, although both catalysts have the same platinum content, the signal in catalyst B is more intense. The nature of catalysts explains this: while catalyst A is a set of clusters, catalyst B is a physical mixture, so surface segregation is impossible. For the synthesis technique, the metal with lower surface energy will tend to segregate into the surface (Brongersma et al. 2010).

Inductively Coupled Plasma Mass Spectroscopy (ICP-MS) is a relatively new unique technique for measuring trace-level elements in liquid samples. Those samples are nebulized to create a fine aerosol transferred to an argon plasma. The high temperature of the plasma atomizes and ionizes the sample generating ions that are extracted at the interface of the equipment and placed on lenses called optical ions. These lenses focus and guide the ion beam to the mass analyzer, where they separate according to their mass/charge ratio (Figure 1) (Wilschefski and Baxter 2019).

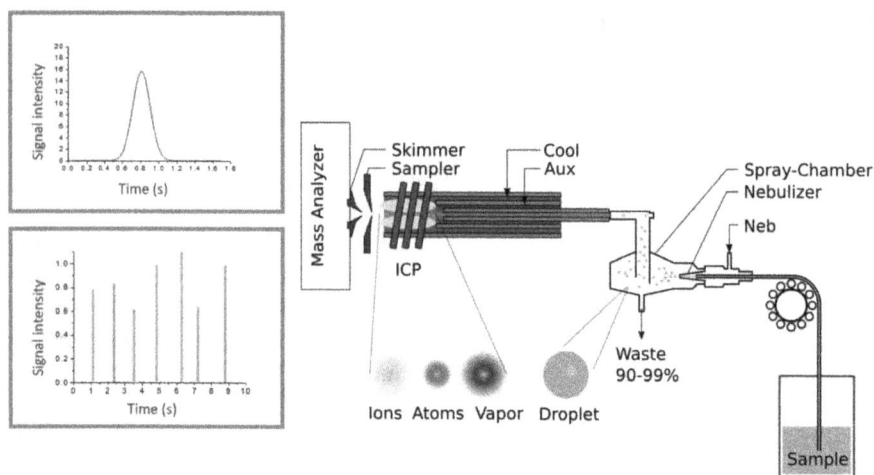

Figure 1. ICP-MS scheme (Tofwerk 2022).

One of the major concerns related to nanomaterials is their effect on human health. ICP-MS has been used to detect synthetic amorphous silicon (SAS) in pyrogenic or precipitated form nanomaterials, which are used as food additives. There is evidence that "pure" state SAS is potentially toxic. To prove this, tissue and blood samples were taken from rats in contact with additive foods and diluted to an adequate concentration through microwave digestion. The analysis showed that compared to tissues that were not in contact with SAS, traces of agglomerates of SiO_2 nanoparticles deposited mainly in the liver were found, so reducing the use of this nanomaterial would be suggested (Aureli et al. 2020).

3.5 Surface Description

The surface description refers to the composition, structure, electronic charge, and surface area of the nanomaterial, which significantly influence the material's functionality (ASTM 2019).

Fourier-Transform Infrared Spectroscopy (FTIR) is an electronic technique used to obtain a spectrum of absorption or emission of light in the mid-infrared (wavelength range between 4,000 and 400 cm^{-1}). The chemical fingerprint of the nanomaterial surface is generated. The sample can be solid, either mixed with KBr (optically transparent in the light of the technique) or complete, using an ATR accessory. Samples can also be measured in suspension or dissolved in different solvents. The difference with infrared spectroscopy (IR) is that the signal is treated to obtain the spectrum.

Nanomaterials embedded in matrices of different natures have diverse complementary effects if compared to the same material added at the micro or macro scale. Recently, bitumen has been mixed with styrene-butadiene-styrene (SBS) rubber as a reuse option for the latter. However, some properties such as plastic deformation or environmental resistance decrease affecting the performance during its application. Different nanoclays begin to be added to improve the SBS-bitumen mix: two compounds based on nanotitania (nano TiO_2) were analyzed with this technique, the first mixed with organically expanded vermiculite and the second with organic montmorillonite. For the compound spectrum with vermiculite, two FTIR peaks were observed at 1024 cm^{-1} and 448 cm^{-1}. They correspond to vermiculite absorption by SBS, ensuring an effect on it. However, no formation of compounds was found between vermiculite and TiO_2. For the montmorillonite compound, an interaction is generated as an interlayer with the bitumen, avoiding an adequate dispersion. This means that the presence of the two compounds in bitumen reduces the reaction of bitumen oxidation and degradation of SBS due to UV light but also improves the performance (Zhang et al. 2022).

Another technique that presents the fingerprint of a nanomaterial is Raman spectroscopy. This analysis uses laser light that hits the material's surface, which absorbs it and changes the internal energy. The energy change results in a higher level of vibrational or rotational energy; when the molecule relaxes, radiation is emitted. This radiation is scattered light captured by the detector but at a different frequency. The detector's sensitivity is greater than FTIR since it must increase or decrease the intensity of the scattered light to give the actual information of the material. In nanotechnology, Raman spectroscopy identifies phases and phase transitions, determines which region is amorphous or crystalline, the defects, and determines the size, shape, and uniformity. Nanomaterials based on ferrite nanocrystals are used in storage devices, recording media, radars, communication systems, biomedicine, water purification, and electronic systems due to their superparamagnetic nature, high surface-to-volume ratio, chemical stability, and non-toxic behavior. Its applicability and final properties depend on the synthesis method, dopant, chemical composition, ionic distribution, particle size, and processing temperature. The effect of sol-gel auto-combustion synthesis was analyzed by Raman spectroscopy for ferrite nanocrystals doped with Mg-Ag-Mn in different concentrations. It was obtained that uni, bi, and three-dimensional vibrations correspond to the spinel and hexagonal ferrite of the nanocomposite. Also, an increase and reduction in the vibration modes were found due to the homogeneous distribution of the dopant ions on the interstitial sites in the main crystal lattice (Jasrotia et al. 2020).

Branauer-Emmett-Teller (BET) surface area analysis is the multipoint measurement of the specific surface area (m^2/g) using the absorption of an inert gas, usually nitrogen, which flows continuously over the solid, dry sample. Due to van der Waals forces, the surface of the nanomaterial forms a gas monolayer, which is recorded along with the absorption rate. Both calculate the superficial area and its porous geometry (Meretics 2022). Also, its limits are mobile and increasingly permeate different areas. Metal-organic framework (MOF) is a class of hybrid materials made up of metal ions and organic ligands that are periodically assembled and used as support materials. Its main characteristic, which shares with nanotechnology, is the ability to focus on various frontier fields (Liu et al. 2018); so what would happen if a nanomaterial-based on MOF is made? This was answered in 2018 when a Cu-MOF@Carbon nanocomposite was studied to catalyze the decomposition of ammonium perchlorate (AP). It provides the oxygen needed for propellant combustion in addition to releasing energy during its decomposition to promote the combustion. The pore parameters of the nanocomposite are a critical factor for its performance as an AP catalyst. The BET analysis showed the presence of meso and micropores with a specific surface area of 10.27 m^2/g. The pore volume found shows the binding of carbon with Cu-MOF at the surface, and this synergistic effect positively influences the thermal decomposition process of AP (Wang et al. 2018).

Secondary Ion Mass Spectroscopy (SIMS) is one of this chapter's few destructive analysis techniques. A layer of material is removed through a beam sputtering ion in the vacuum that provides enough energy to generate positive (e.g., carbon) and negative (e.g., oxygen) ions from most elements (Figure 2). They are analyzed by mass spectrometers giving a surface characterization of the nanomaterial. Its main disadvantages are that some elements cannot be analyzed on specific substrates or matrixes, and instrumentation is very expensive (Montana State University, 2022). For certain industrial applications, the performance stability of nanomaterials is critical as they should be more durable in application than their macromolecular counterparts. Perovskite has begun to be used in solar cells, trying to replace traditional ones; however, its stability has limited the massification of the technology. SIMS in Time of Flight (TOF) mode (three-dimensional analysis) combined with XPS was used to quantitatively characterize the effect of aging in solar cells. Three solar cell architectures were analyzed to investigate the mechanisms of degradation. The reconstruction of two-dimensional images shows that after 24 h of sun exposure leads to partial degradation of the interface, but not homogeneously; it decreases the performance (Figure 2). The authors recommended substrate modifications with graphene to improve electron transfer and total solar cell stability (Busby et al. 2018).

Nuclear magnetic resonance (NMR) is one of the most widely used techniques to determine the molecular structure attached to the surface and the size of nanoparticles in the solution. Through the application of a magnetic field, the interaction of the

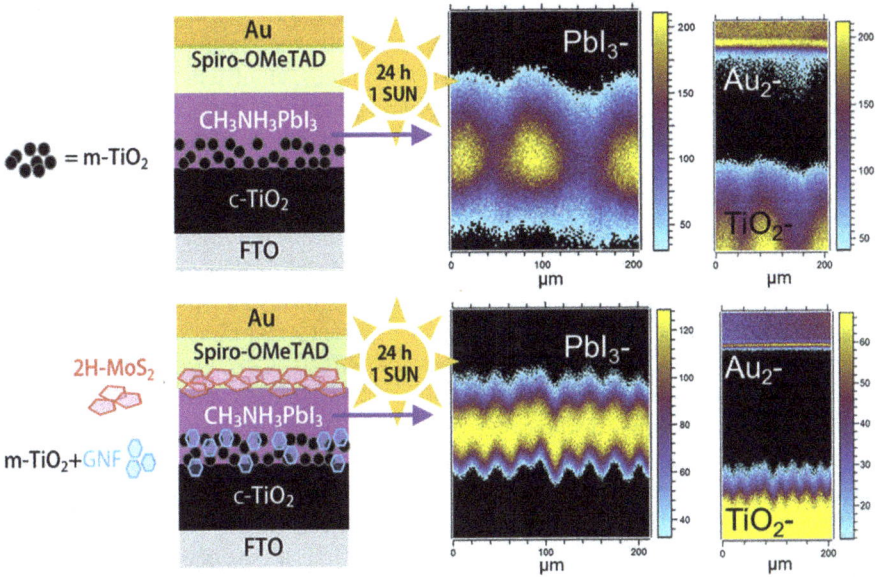

Figure 2. 3D SIMS analysis of new (a–c) and 24-hour aged (d–f) solar cells. The decrease in the PbI_3 signal suggests a progressive decomposition of the perovskite absorber material (Busby et al. 2018).

nuclear spin of the surface sample is observed and measured, obtaining a spectrum of the atomic groups for a specific atom. For this technique, it is necessary to have previous information about the tested material to understand the spectra properly. Perovskite quantum dots were analyzed for use in optoelectronic materials. The properties in the application depend on the surface chemistry, so NMR, in this case, determines the aging of nanomaterial and ligand exchange. The presence of oleic acid was obtained. With the information obtained by NMR, the ligand density of the bound oleate was determined, and according to theoretical values, there is an incipient formation of ligands while under the presence of 10-undecenylphosphonate, the ligand density increases. This is sufficient to determine an appropriate surface alteration according to the application needs (Chen et al. 2020).

3.6 Aggregation

This feature describes how a set of nanomaterials interacts by separating into agglomerates or aggregates. The difference between the two associations is in the resistance to staying together. This characteristic does not apply to all types of nanomaterials, as certain groups must keep each "nano-object" separate to achieve the objective of their application (ASTM 2019).

Zeta Potential analysis is a technique for determining the surface charge of nanoparticles in colloidal solutions. In suspension, each particle has a thin layer of ions with an opposite charge and a subsequent layer of loosely associated ions, allowing the particle to move through the medium. The zeta potential is the voltage at the edge of the layer plane where there is no longer an association with the particle. It is considered to determine the stability of suspended nanoparticles and is the main factor in the absorption of nanoparticles in cell membranes when used in biomedical applications. This technique measures the magnitude of the electrostatic charge between particles giving details regarding dispersion, formation of aggregates, and flocculation with which the formulation of dispersions, emulsions, and suspensions can be improved. For example, if two adjacent particles have zeta potentials of the same sign, they will not agglomerate but may also have a chemical nature that favors agglomeration. The suspension is measured in a folded capillary cell with two electrodes through which the instrument applies voltage. A HeNe laser illuminates the suspension giving the value of electrophoretic mobility (nanoComposix 2022). Continuing with the environmental effect of nanomaterials during and after application, it has been found that especially metal nanoparticles in marine environments tend to interact with cells of different organisms, forming colloids (nano-bio interfaces) that can generate adverse effects. CeO_2 nanoparticles began to be used in catalysts, sunscreens, fuel cells, fuel additives, and biomedicine, so their potential polluting effect has been analyzed. *In vitro* analyses were performed using primary hemocytes from *Mytilus galloprovincialis* (Mediterranean mussels) and observing their response to the presence of these nanoparticles in different media (ultrapure water, artificial seawater, and filtered hemolymph serum). After some time, in artificial seawater, there was an increase in hydrodynamic radius showing agglomerations which did not occur in serum and ultrapure water. The zeta potential changes from positive in ultrapure water to negative in serum and ultrapure water. It was also observed, for nanoparticles manufactured by different methods, that depending on the pH during its manufacture, there are changes in the potential zeta and, therefore, the capacity of biocrown formation, which has a negative impact on hemocytes (Sendra et al. 2018).

3.7 Interfaces

This characteristic defines the boundaries between different regions in the nanomaterial, if applicable. The interface can be described by its location, the regions that define it, the area of the boundary, or the type of interaction that occurs in it (ASTM 2019).

Electron Energy Loss Spectrometry (EELS) analyzes the energy distribution of electrons emerging from a thin sample due to inelastic interactions. The resulting spectrum contains information on atoms and electronic structure, bonds, distribution, coordination number, dielectric constant, and band gaps. Commonly, it is coupled to TEM, where the interaction of its beam is used to obtain the compositional information of deep nuclei (core loss) or the optical information through the excitation into low-lying states (Hachtel et al. 2018). One of the great advantages of this technique is that it allows the measurement in blends or amorphous nanomaterials using the areal density of specific elements. The number of incident electrons is obtained using the core, EELS, and conversion of the spectrum counts using a calibrated charge coupled device (CCD) (Tian et al. 2019).

3.8 Topology

The topology describes the connectivity and continuity of the nanomaterial or its components, including its relative position (ASTM 2019). In addition to SEM, TEM, and AFM, surface topology can be studied using 3D profilometry. Although this technique is typically related to macro-scale surface analysis, it began to be used to characterize the roughness, topography, and flatness of nanomaterial surfaces. Unlike the physical probe used on bulk surfaces, an optical profilometer is used. There are different techniques, but all use a light source directed to the sample, and a detector or camera generates the 3D surface. They can also have accessories to detect phases and even project patterns onto the surface for quality control by comparison (NanoScience Instruments 2022).

The study of surfaces, contact, and lubrication, and tribology, has also been positively affected by nanotechnology since different types of lubricants can be custom designed for applications where conventional ones are not allowed or do not meet the needs. For example, conventional motor oil can be added with nanotubes and nanocomposites to reduce friction and wear, decreasing maintenance and positively affecting the emissions of machines. Comparing the topology of an oil-lubricated bronze with bronze topologies lubricated with oils with multi-walled carbon nanotubes and zinc oxide nanoparticles revealed a significant reduction in wear when the nanotubes are combined with nanoparticles, which allow the use of these additive lubricants in machine elements such as gears and bearings (Ajay Vardhaman et al. 2020).

4. Thermal Properties

The thermal properties of nanomaterials depend on many factors, including analyzed temperature gradient, physical and chemical properties of the material, and direction of heat conduction, among others.

4.1 Thermal Conductivity

Thermal conductivity is the most used property to quantify heat transport through the material when analyzing a thermal process. It is known that heat is conducted in a material according to the atomic structure. In a nanomaterial, the thermal attributes are different from those observed at the macromolecular scale, and therefore wholly different techniques are necessary. The Laser Flash technique (LFT, LFA) employs a xenon flash lamp that heats the sample at one end. The temperature increase is recorded in an infrared detector, and a data acquisition system generates a temperature vs. time curve to give the thermal conductivity. Phase change materials have a high thermal storage capacity and can keep their temperature constant during a phase change, so they are used in solar energy storage systems; however, they have a low thermal conductivity conducing to a prolonged process of storage and release of energy; however, the introduction of nanoparticles at concentrations as low as 2 wt% can generate a significant change in this property. The palmitic acid was loaded with nitrogen-doped graphene in different mass fractions between 1 and 5 wt%. A 500% increase in the thermal conductivity was obtained for the maximum concentration without affecting its reliability and chemical durability (Mehrali et al. 2014).

4.2 Thermal Diffusivity

Thermal diffusivity is a material property that governs the process of thermal diffusion over time. This property is directly related to its density, thermal conductivity, and specific heat at constant pressure. A nanomaterial that possesses high diffusivity means that it quickly transfers heat.

The thermal Lens Technique (TLT) measures the temperature increase in an illuminated sample due to the non-radiative relaxation of the energy absorbed from a laser. It is a highly-sensitivity technique applicable mainly in analyzing new nanomaterials whose thermal properties are critical in the applications. For nanomaterials, thermal diffusivity depends on size, so it is necessary to find the appropriate value where this property is magnified. The diffusivity of CdTe colloidal quantum with different sizes was studied in the dual-mode of the TLT to find its lowest value and be able to be applied in thermal insulators. It was observed that the sample with 60 minutes in a pressure vessel at 80°C had the lowest thermal diffusivity. It was determined that this value decreased further if the solution of $CdCl_2$, thiomalic acid, and Na_2TeO_3 used for synthesis had a concentration of 0.05 mg·mL^{-1} (Nideep et al., 2020).

4.3 Specific Heat

The specific heat represents the energy required to change the temperature of a unit mass of material by one degree of temperature. The specific heat represents the material energy per unit mass at a specific temperature. Specific heat can be measured at constant pressures, Cp, and constant volume, Cv. Similar to thermal diffusivity, in nanomaterials, the value of this property changes with size but also with shape. In general, the specific heat increases with the size of the nanomaterial and decreases as follows: nanoparticle > nanowire > nanofilm (Figure 3).

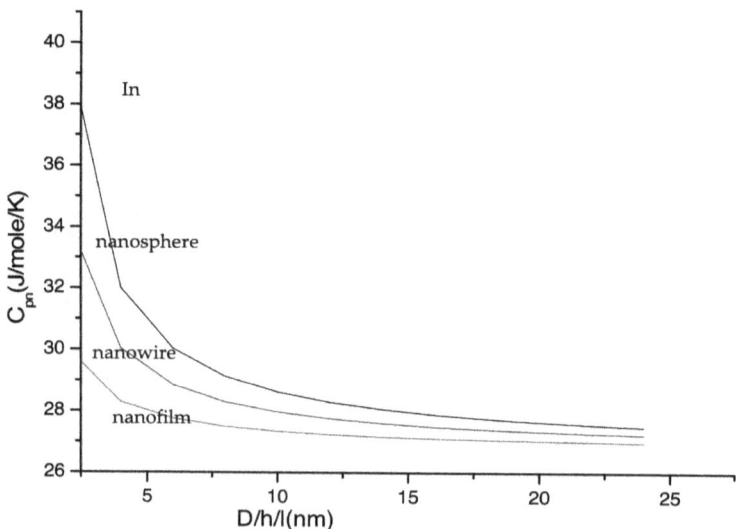

Figure 3. Variation of the specific heat of In (Indium) nanomaterials with size and shape (Singh et al., 2017).

The preferred technique for measuring specific heat is Differential Scanning Calorimetry (DSC), with which other thermal properties such as melting temperature, crystallization temperature, glass transition temperature, precipitation, curing mechanisms, and phase changes can also be detected. This technique uses small sample sizes (between 1 mg and 20 mg), making it ideal for handling nanomaterials. In a sample, the temperature is increased at a constant velocity, and the response is compared with a reference. It identifies the processes in terms of change of slope of the curve or generation of concave and convex peaks that, depending on the configuration of the data acquisition system, can be considered endothermic processes (absorb heat) and exothermic (release heat). Through DSC, the specific heat during the combustion of carbon nanoflakes of different thicknesses was analyzed, heating up to 1273 K at 10 K·min^{-1}. It was observed that the specific heat increases with the synthesis time since more layers of graphite are formed. However, over time defects such as pores or incorrect superposition of layers also increase, so this behavior cannot be entirely attributed to the time variable (Arkhipova et al. 2020).

4.4 Thermal Expansion Coefficient

The coefficient of thermal expansion is related to the volume changes in nanomaterials due to temperature variations. As in the previous thermal properties, the coefficient of thermal expansion depends on the shape and size for most nanomaterials. The volumetric expandability decreases with the increasing size of the nanomaterial while decreasing as follows: nanoparticle > nanowire > nanofilm (Goyal and Gupta 2018).

Dilatometry (DIL) is a technique in which a dimensional change of the nanomaterial is measured, whether linear, surface or volumetric, as a function of temperature under a negligible load. This technique can characterize phase changes, density changes, decomposition, and controlled sintering. The sample is placed in a holder under a controlled atmosphere where the temperature increases progressively, and the dimensional change is detected. This technique uses a liquid thermometer for temperature recording as it is more sensitive to changes. At the end of the test, a curve is obtained that records the expansion rate as a function of temperature. Electromobility has become an option for reducing emissions that affect the environment since lithium batteries are used instead of petroleum-based fuels. The lifetime of these batteries is related to the dimensional stability of the electrodes during the charging and discharging process. Since the movement of lithium ions causes expansion and contraction, crystalline distortion and phase transformations are observed. The presence of carbon nanotubes in lithium transition metal oxide electrodes affects changes in electrode thickness and height during the charge and discharge cycle, demonstrating an 87% reduction by DIL. Also, the electrode aging causes an increase in the expansion range, losing the cell potential. Nanotubes reduce both situations, and as a side effect, they increase the usable voltage range of the battery (Mousavihashemi et al. 2022).

In addition to the properties mentioned above, other techniques thermally characterize any nanomaterial efficiently. Thermogravimetric analysis (TGA) measures changes in the mass of a sample as the temperature increases since the test is performed on an analytical balance in an inert atmosphere. In nanotechnology, TGA is used to determine the amount of coating on the surface of a nanomaterial, study vaporization or decomposition processes used in synthesis, nanoarray stability in a matrix, and detect impurities in nanomaterials used in biomedical applications.

Pyrolysis has become a widely used strategy to reduce waste in a controlled manner and to obtain by-products that can be useful or disposable in a safer way. Cereal husks are biological wastes, and with other types of biomasses, it has been shown that under controlled conditions, they become a precursor of carbon nanotubes using catalysis in a microwave-heated oven. Through TGA, by-products can be determined in the catalysis of nanotubes and considered impurities that affect their performance and properties. Analyzing catalysis, the presence of iron oxide favors the early decomposition of the nanotube. It is important to clarify that for this technique, it is necessary to have a good knowledge of the background of the material to be able to identify the behavior of the resulting curve (Asnawi et al., 2018).

5. Rheological Properties

Viscosity is the most widely used material parameter to determine the flow behavior of nanomaterials during processing, synthesis, or application. Rheological properties are sensitive to the structure, concentration, particle size, shape, and surface characteristics of nanomaterials embedded in a suspension. The flow behavior of these nanofluids is expressed in terms of shear stress and shear deformation rate and is described as Newtonian behavior (directly proportional relationship between stress and rate) or non-Newtonian (non-direct relationship). The presence of agglomerates and aggregates also affects rheological behavior, so it is necessary to understand it before making simplistic assumptions (Woodhead Publishing 2017). The characterization technique uses a rheometer, and depending on the information obtained from the nanomaterial, a different type of rheometer is used. In applications related to heat transfer, for example, refrigerant liquids, nanofluids have taken relevance because the high surface area favors thermal conductivity and subsequent cooling. However, the formation of agglomerates, generation of liquid layering at the particle-fluid interface, thermophoretic effects, or Brownian movements of low extension affect the performance negatively. The study of ethylene glycol nanofluids with different carbon-based nanomaterials was carried out in a rotational rheometer with a double cone in a strain rate range between 1 and 1000 s^{-1} to obtain the storage module, loss module, and complex viscosities depending on the deformation. All nanofluids were found to have a viscosity that decreases nonlinearity with an increasing deformation rate; a non-Newtonian behavior, called thinning shear or pseudoplastic, is typically observed in molten thermoplastic polymers. Behavior at low deformation rates demonstrates the presence of agglomerates that break as the deformation rate increases. A plateau stabilization is in the region of interest for the nanofluids' application (100 and 1000 s^{-1}). The behavior is also explained by the difficulty of the remaining agglomerates' orientation in the flow direction. It is also clear that the nature of the embedded nanoparticle affects the behavior in the early stages; the carbon black (CB) nanofluid showed the highest viscosity, which coincides with the expected behavior since it has a high tendency to form aggregates and agglomerates while the nanofluid with lower viscosity corresponds to that of graphene nanoplatelets (Vallejo et al. 2019).

Surface tension is another rheological property that shows the intensity of contact between two phases, usually solid-liquid, and is important for applications such as coatings, paints, and dispersions. It is generally used to measure the contact angle, which is a direct measurement of the tangent angle at the three-phase contact point of a nanofluid droplet using a goniometer telescope. Membrane manufacturing is another application that has been widely penetrated by nanotechnology. Ultrafiltration uses cellulose acetate polymer membranes modified with graphene oxide nanosheets and MoS_2 to change the surface tension and allow water purification. High hydrophilicity means a lower contact angle allowing greater water permeation and efficiency. The graphene oxide nanocomposite showed the lowest contact angle compared to the MoS_2 nanocomposite, and its manufacture also generates greater porosity, improving filtering efficiency (Vetrivel et al. 2018).

6. Mechanical Properties

The mechanical properties of nanomaterials and other properties presented above differ considerably from bulk behaviors due to the high surface area/volume ratio. For a successful application of the nanomaterial, it is necessary to understand the mechanical properties such as hardness, elastic modulus, and interfacial adhesion, in addition to the effect of size and shape. As a generality, toughness, and hardness are reduced with the content of the nanomaterial because its presence increases the growth of the crystals and favors the development of fractures; hence, they are usually used in very low proportions (Wu et al. 2020).

Different characterization techniques focus on determining the mechanical properties in nanomaterials or the effect in bulk matrices. Nanoindentation is a technique widely known in industry and academia that mechanically characterizes thin surfaces, but in nanocompounds allows for determining the effect of nanomaterials by configuring the test to a certain percentage of penetration. This technique, similar to the surface hardness test, uses a nanoindenter in the form of a diamond pyramid. Depending on the desired information, a point value of nanohardness or a charge-depth curve in a charge-discharge cycle can be given where various mechanical properties can be calculated indirectly. The biomedical area has benefited from the presence of nanomaterials because it has allowed not only to "miniaturize" elements but also to facilitate the design of nanocomposites according to specific needs. In dental treatments, it is necessary to know the mechanical performance of the materials, mainly for restorations, due to the high static and dynamic loads they must withstand. Nanocomposites and nanoionomers have been developed to replace conventional restorative materials for their best general properties and biocompatibility. Through a nanoindentation curve, the hardness and Young's modulus of two conventional materials (microhybrid resin and glass ionomer) were compared with a nanohybrid compound and a nanoionomer. It was observed that the hardness of the nanocomposite is the highest, followed by the microhybrid compound, the nanoionomer, and the ionomer. This behavior is also found in the elastic modulus, yield stress, and fracture toughness. The nanocomposite can successfully replace any restorative material, but it is necessary to rethink the nanoionomer processing since the acid used during synthesis reacts with the nanoparticles lowering the final properties (Peskersoy and Culha 2017).

Mechanical Dynamic Analysis (DMA) is a technique that determines the kinetic properties of a material using the change of deformation or stress during several cycles or oscillations over time. During the measurement, a sinusoidal force is applied, and the deformation caused is detected (and vice versa) at a constant temperature or by changing it over time. Sinusoidal force can be tension, compression, or shear, according to the need for analysis. In aerospace applications, nanomaterials broke the invisible ceiling of the mechanical performance of already outstanding materials and composites with the presence of nanomaterials. Carbon fibers embedded in polymer matrices have the "slightest" disadvantage of poor bonding, but with surface modifications using nanoparticles, the wettability of carbon fiber can be changed. DMA was used to determine the effect of coating carbon fiber with nickel nanoparticles on polymer matrix compounds. The loss module (ability to dissipate mechanical energy in the form of heat), the loss tangent (dissipated energy), and the variation of the glass transition temperature with the coating time were analyzed. For coating times up to 20 minutes, a reduction in the loss modulus was observed due to increased friction between the coating nanoparticles and the matrix materials. After 20 minutes, the size, density, and distribution play against it, confirmed by the curve of the loss module with the displacement module. The loss tangent decreases consistently, indicating a decrease in damping. It is clear that 5 minutes of coating is not enough time to affect the mechanical properties of the compound, but in general, coating with nickel nanoparticles favors mechanical interlocking between the matrix and fiber (Yadav et al. 2018).

7. Magnetic Properties

These properties will be related to nanomaterials based on nickel, iron, cobalt, chromium, manganese, and gadolinium, among others, exhibiting ferrous and ferrimagnetic behavior, that is, responding to the presence of a magnetic field. In addition to the system's nature, the nanomaterial's magnetic properties depend on the synthesis method. Also, due to the size of the nanomaterial, a phenomenon called superparamagnetism can occur where a thermal fluctuation can spontaneously demagnetize which instead of being considered a disadvantage, makes them perfect candidates for therapy and targeted drug delivery. Another effect that relates magnetism and temperature in nanomaterials is the magnetocaloric effect, where the material is heated in a magnetic field. Due to their high surface area, nanomaterials exchange heat more efficiently and can also be used in hyperthermic therapies (Rikken et al. 2014).

SQUID (Superconducting Quantum Interference Device) is an electronic technique used to generate a mapping of magnetic fields above sample surfaces. It is a measurement of high sensitivity but of low spatial resolution. A small SQUID is mounted on a tip that scans the sample's surface to be analyzed and detects the voltage levels generating intensity images. Due to the rapid development of nanomaterials, it has not achieved its effect on human and animal health in some applications. As previously indicated, magnetic particles have found a niche in targeted drug delivery. However, depending on the microcellular environment where they are used, their stability is affected, generating intraendosomal degradation, which is an iron overload (intracellular toxicity) that can be dangerous in organisms with moderate iron metabolism. The magnetometry SQUID was used in different biological samples (liver, left ventricle, kidneys, aorta, and arterial blood) to detect possible agglomerations of iron oxide nanoparticles and differentiate them from the biogenic iron that is always present in any biological system. Control tissue was compared with a tissue in contact with the iron oxide targeted drug delivery system to which the magnetization mass was measured in terms of the magnetic field, finding that there was no difference between the tissues (the technique under this condition does not differentiate between iron biogenic and iron oxide nanoparticle). However, in the curve that relates the mass of magnetization with temperature, it is observed, for the tissue contact with the delivery system, that instead of finding a supramagnetism at high temperatures, an increase in the mass of magnetization appears, which confirms the presence of iron oxide nanoparticles. The highest levels were found in the blood, followed by the aorta, liver, ventricle, and kidneys (Škrátek et al. 2020).

Vibrating Sample Magnetometer (VSM) is a technique for measuring the magnetic properties of materials as a magnetic field function using Faraday's law of induction. The sample is located inside a constant magnetic field and observes its dipole alignment and the electric current generated. Typically, a magnetic hysteresis curve is obtained. Some synthetic pigments have harmful effects on both the environment and the human body. Methylene blue in high concentrations causes breathing difficulties, nausea, mental dysfunction, and methemoglobinemia. Fortunately, nanotechnology has served for photocatalytic purification with advanced oxidation processes, effectively removing the compounds producing hydroxyl radical and non-selectively degrading these organic compounds. The hybrid nanomaterial $LaFeO_3 \cdot Fe_3O_4@SiO_2\text{-}NH_2/PW_{12}$ has shown excellent efficiency when degrading methylene blue and methyl orange, but in addition, it was magnetically analyzed to determine its possible recovery. From the curve of variation of the magnetization with the magnetic field, a hysteresis cycle with paramagnetic properties was observed. An analysis was also carried out that reproduced the separation of the hybrid nanomaterial from a solution using an external magnet. It is observed in the hysteresis cycle that the magnetization is recovered easily from the solution (Mahmoudi et al. 2020).

Mössbauer spectroscopy is a technique widely used to study magnetic particles from a magnetic relaxation spectrum obtained by radiation with a low-energy gamma-ray beam. A detector measures the intensity of the beam transmitted through the sample. The atoms of the emission source must be the same isotopes of the atoms as in the sample that absorbs them. A disadvantage of this technique is the difficulty of spectrum analysis, and each change in transmittance with respect to speed can mean countless reasons.

In some cases, a nanomaterial is suitable for a specific application, but the synthesis is not adequate to scale it at an industrial level. Strontium hexaferrites are nanomaterials that have shown effectiveness in microwave absorption, recording in magnetic media, signal processing, telecommunications, and audio systems, among others. However, the synthesis techniques do not ensure dopants' preferential occupation. A new synthesis with manganese was presented to obtain more reliable materials. Mössbauer spectroscopy was used at room temperature giving information about the occupation of preferred sites of dopants in the lattice (Almessiere et al. 2019).

Magnetic Force Microscopy (MFM) is a relatively new method for mapping the distribution of magnetic field and different kinds of magnetic interactions using magnetic forces or a gradient of magnetic forces on the surface of a sample. A magnetic tip is used in a cantilever very close to the surface, and the displacement of the tip is detected optically (Figure 4). At first glance, this technique is the same as AFM; however, MFM differs from AFM in that its tip does not make direct contact with the sample and is at an approximate distance of 30 nm.

Nanomaterials based on manganese oxides have become of great interest because they have various chemical-physical properties, are economical, environmentally friendly, polymorphic, and structurally flexible. To further manipulate these characteristics, ionic dopants are used. The use of fluorine as a dopant is studied since it generates thermodynamically stable nanomaterials, especially for mono and two-dimensional nanomaterials (nanorods and nanofilms). A parallel analysis was performed using AFM and MFM to obtain the topography and magnetic properties of the magnetic domains for different growth temperatures of the nanomaterial (Figure 5). At a higher temperature, more defined structures (nanorods) with highly porous morphology were found. Magnetic polarization is also reflected in the upper part of the nanoaggregates generating magnetic anisotropy, which can be used in high-density magnetic recording media (Barreca et al. 2018).

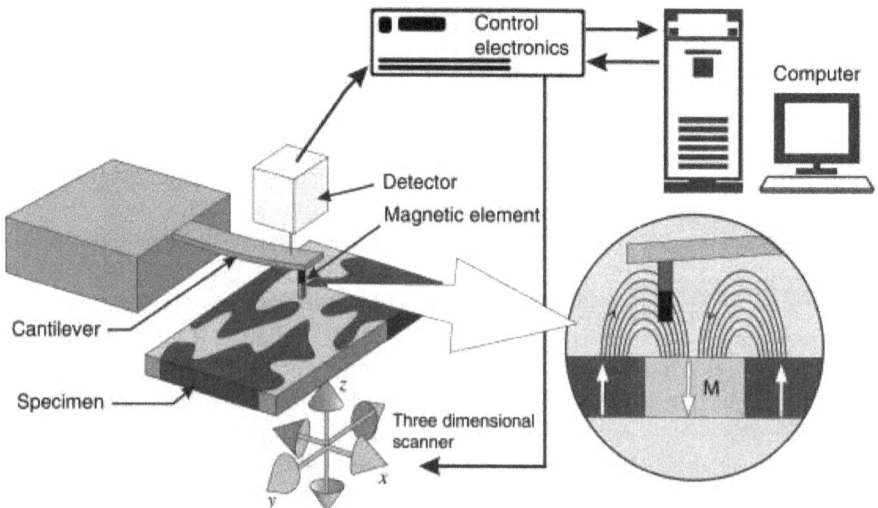

Figure 4. Diagram of the magnetic force microscopy (Abelmann 2017).

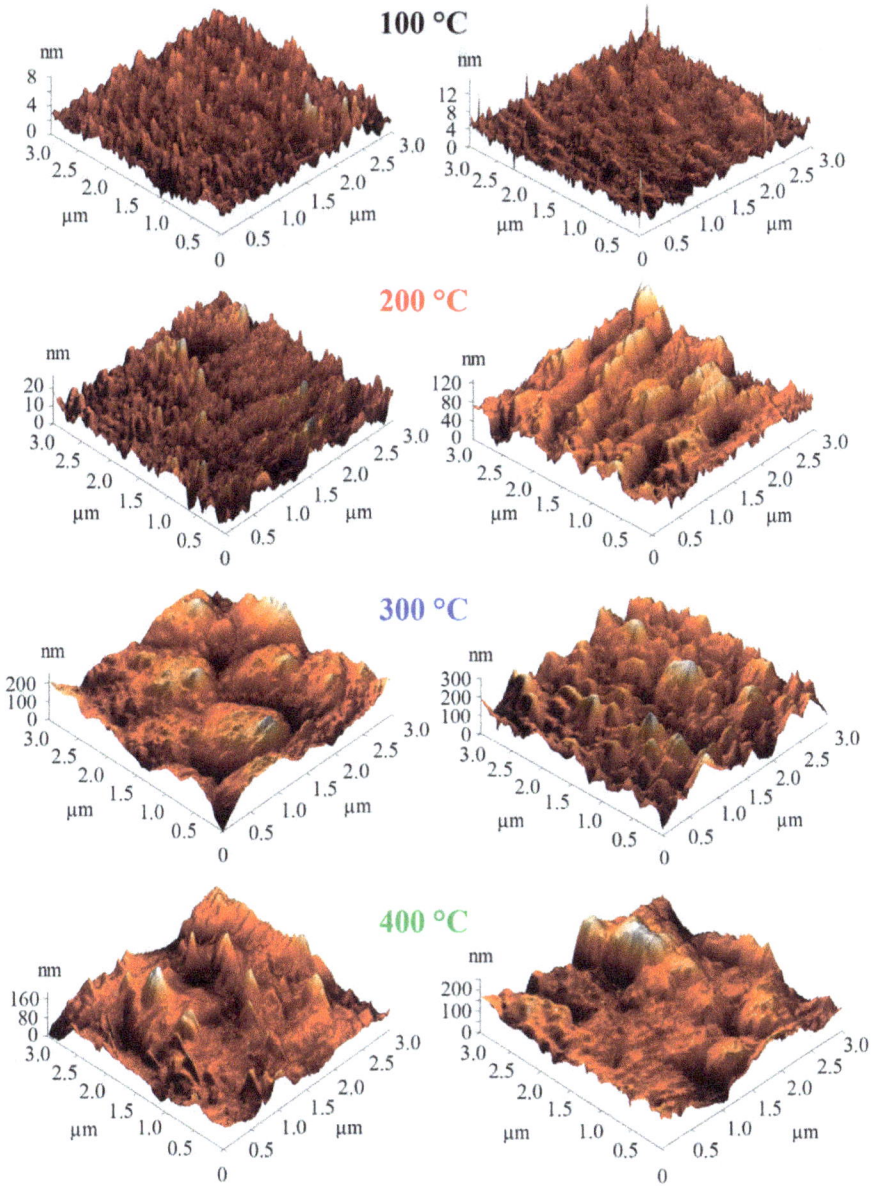

Figure 5. Topography (left) and MFM chromography (right) of manganese oxide samples obtained at different growth temperatures (Barreca et al. 2018).

Ferromagnetic Resonance Spectroscopy (FMR) is a technique used to test magnetic nanomaterials' magnetization dynamics. This technique has two methods: the sample is placed in a resonant cavity, which is illuminated with microwave radiation at a fixed frequency, recording the magnetic field, and, in the sample, electrical measurements are made based on strip lines and coplanar waveguides. The latter method is preferred for nanofilm analysis. This technique is known as Electron Paramagnetic Resonance (EPR) when analyzing low magnetic fields. EPR has been applied to elucidate the magnetic properties in applications such as gas sensing, supramagnetic purification of wastewater, and removal of radionuclides. In zinc ferrites doped with cerium ions at different concentrations, it has been analyzed how their magnetic anisotropy and dipole interactions change. For one concentration, a minimum value was obtained for the width of the resonance line, which means better uniformity. Under the other concentrations evaluated, the decrease in dipole interactions and superexchange was found, making these ferrites candidates for microwave absorption materials (Alshahrani et al. 2021).

8. Electrical Properties

These properties are related to the material's behavior in the presence of an electric current or voltage. Given the broad nature of nanomaterials, they cannot be categorized only into conductive or insulating materials because it depends on the type of atoms, the shape, and size, which does not occur in bulk. The characteristics are electrical conductivity, coefficient, and electrical polarization.

Electrochemical Impedance Spectroscopy (EIS) is a rapid, non-destructive technique for detecting the electrical properties of nanomaterials. It consists of applying a sinusoidal voltage and measuring the response current. A Faradaic impedance spectrum called Nyquist Plot is obtained, and it contains information on capacitance, the ohmic resistance of electrolytic solutions, resistance to electron transfer, and Warburg diffusion impedance. This spectrum can be considered the electrical fingerprint of the nanomaterial. Graphene coatings have become an interesting and effective corrosion barrier; however, their long-term performance eventually generates galvanic corrosion. A new hybrid graphene oxide-ionic liquid (IL) nanomaterial has been designed. EIS is used to check the anticorrosive capacity of the new material, that is, to observe the inhibitory effect of the ionic liquid and compare aged steel surfaces (at different times) with pure epoxy, graphene oxide, and the new nanomaterial with two different concentrations of graphene oxide. The Nyquist and Bode diagrams showed a decrease in the capacitive arc, which shows a decrease in anticorrosive capacity. For surfaces with pure epoxy and graphene oxide, only arc loss is observed even on the first day of aging, while the new material is observed up to 10 days later. The presence of two arcs on pure epoxy and graphene oxide surfaces indicates that the coating was saturated with the electrolytic solution. The decrease of graphene oxide concentration increases the impedance modulus, showing that the protective effect is directly related to the ionic liquid (Liu et al. 2018).

9. Optical Properties

The optical properties of the nanomaterials, as the previous properties analyzed, are entirely different from their bulk. Light absorption, transmission, reflection, and emission depend on the shape, size, surface characteristics, and internal electronic structure. The color observed in nanomaterials results from the resonance of the wavelengths of light at the time of contact with the outer electronic band. Hence, depending on the size, the nanomaterials of the same nature do not have the same color. Usually, the optical properties are determined through a spectrophotometer where the sample is illuminated, and a detector measures the reflected or transmitted radiation. Ultraviolet-Visible (UV-Vis) and photoluminescence (PL) spectroscopy are sensitive to electronic transitions. In both, the sample is irradiated with a beam of light of a specific wavelength. In the first technique, the detector measures the amount of light absorbed by the sample, while in the second technique, the photons emitted by the sample are detected. Figure 6 shows the UV-Vis absorption spectrum

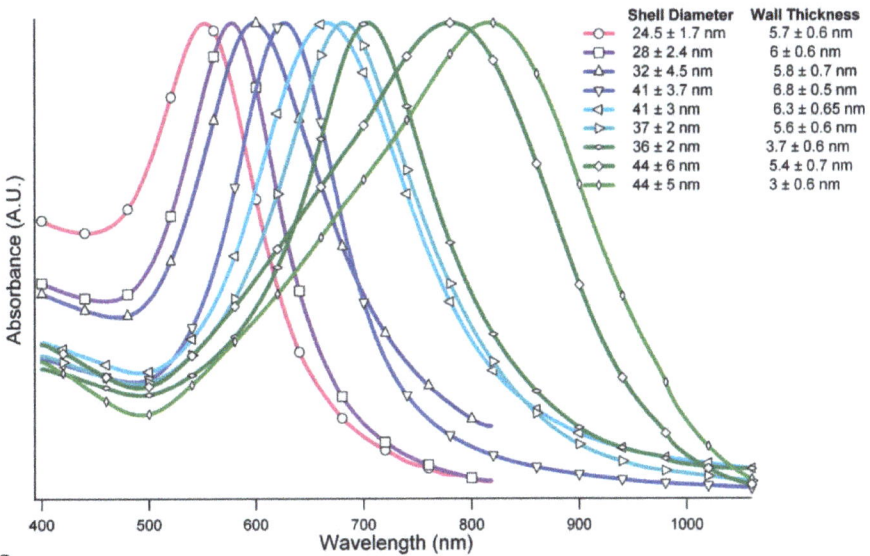

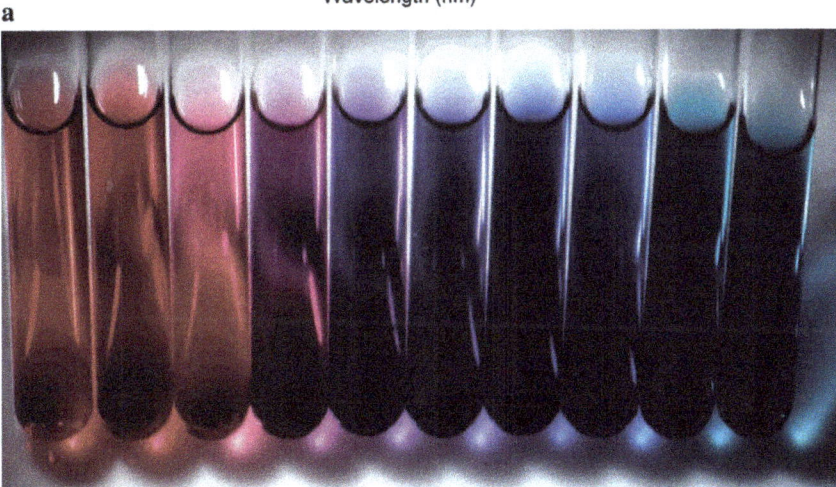

Figure 6. UV-Vis spectrum of nine samples of hollow gold nanoparticles with different diameters and thicknesses complemented by the color of the solutions (Schwartzberg et al. 2006).

of nine hollow nanospheres with different sizes and wall thicknesses. It is observed that the change of the peak wavelength intensity is related to a different color in its dispersion (Schwartzberg et al. 2006). The photoluminescence spectrum of $NaYF_4$ nanoparticles doped with lanthanides alters the optical properties by controlling the nature, level, and the ratio of dopants (Wang and Liu). In nanoparticles, the optical phenomenon is typically not linear, so its use in light energy conversion, detection and sensing, and optoelectronics is very popular.

Final Remarks

This chapter shows the different techniques used to characterize and analyze the properties of different nanomaterials, showing each equipment's principles and operating schemes and an example of the application of each technique. It is clear that no single technique gives complete information about a particular problem, and therefore, they should be supplemented with other techniques as needed. It is always suggested that before analyzing a nanomaterial, carry out a bibliographic search on the best techniques to solve the problem in particular. The full bibliography of this chapter can be of great help to this.

Acknowledgments

The author acknowledges ANID for financial support of the project FONDECYT Iniciación No. 11190596.

References

Abelmann, L. (2017). Magnetic force Microscopy. pp. 675–684. *In*: Lindon, J.C., Tranter, G.E. and Koppenaal, D.W. (eds.). *Encyclopedia of Spectroscopy and Spectrometry*. Academic Press.

Ajay Vardhaman, B.S., Amarnath, M., Ramkumar, J. and Mondal, K. (2020). Enhanced tribological performances of zinc oxide/MWCNTs hybrid nanomaterials as the effective lubricant additive in engine oil. *Materials Chemistry and Physics*, 253: 123447.

Almessiere, M., Slimani, Y., Güngünes, H., Baykal, A., Trukhanov, S. and Trukhanov, A. (2019). Manganese/Yttrium codoped strontium nanohexaferrites: Evaluation of magnetic susceptibility and mossbauer spectra. *Nanomaterials*, 9(1): 24.

Alshahrani, B., ElSaeedy, H.I., Fasres, S., Korna, A.H., Yakout, H.A., Abdel Maksoud, M.I., Amer Fahim, R., Gobara, M. and Ashour, A.H. (2021). The effect of Ce3+ doping on structural, optical, ferromagnetic resonance, and magnetic properties of ZnFe2O4 nanoparticles. *Journal of Materials Science: Materials in Electronics*, 32: 780–797.

Amaro-Gahete, J., Benítez, A., Otero, R., Esquivel, D., Jiménez-Sanchidrián, C., Morales, J., Caballero, A. and Romero-Salguero, F.J. (2019, January 26). A comparative study of particle size distribution of graphene nanosheets synthesized by an ultrasound-assisted method. *Nanomaterials*, 9(2): 152.

Analytik. (2022, May 22). *Analytik*. From Particle Size Characterization by Differential Centrifugal Sedimentation (DSC): https://analytik.co.uk/particle-size-characterisation-by-dcs/.

Anton Paar. (2022, May 14). *Anton Paar Wiki*. From The Principles of Dynamic Light Scattering: https://wiki.anton-paar.com/it-it/i-principi-della-diffrazione-dinamica-della-luce.

Anton Paar. (2022, May 17). *Microwave-assisted Sample Preparation*. From Microwave digestion: Related information: https://wiki.anton-paar.com/en/microwave-assisted-sample-preparation/#:~:text=Acid%20digestion%20is%20the%20most,can%20be%20brought%20into%20solution.

Arkhipova, E.A., Strokova, N.E., Tambovtseva, Y.A., Ivanov, A.S., Chernyak, S.A., Maslakov, K.I., Egorova, T.B., Savilov, S.V. and Lunin, V.V. (2020). Thermophysical study of graphene nanoflakes by differential scanning calorimetry. *Journal of Thermal Analysis and Calorimetry*, 140: 2641–2648.

Asnawi, M., Azhari, S., Hamidon, M.N., Ismali, I. and Helina, I. (2018). Synthesis of carbon nanomaterials from rice husk via microwave oven. *Journal of Nanomaterials*, 2018: 2898326.

ASTM. (2017, August 1). ASTM Standard E2859. *Standard Guide for Size Measurement of Nanoparticles Using Atomic Force Microscopy.* New Conshohocken, PA, USA: ASTM.

ASTM. (2018, January 1). ASTM Standard E2834. *Standard Guide for Measurement of Particle Size Distribution of Nanomaterials in Suspension by Nanoparticle Tracking Analysis (NTA).* New Conshohocken, PA, USA: ASTM.

ASTM. (2019, November 15). ASTM Standard E3206. *Standard Guide for Reporting the Physical and Chemical Characteristics of a Collection of Nano-Objects.* New Conshohocken, PA, USA: ASTM.

ASTM. (2019, September 1). ASTM Standard E3144. *Standard Guide for Reporting the Physical and Chemical Characteristics of Nano-Objects.* New Conshohocken, PA, USA: ASTM.

ASTM. (2021, February 1). ASTM Standard E2490. *Standard Guide for Measurement of Particle Size Distribution of Nanomaterials in Suspension by Photon Correlation Spectroscopy (PCS).* New Conshohocken, PA, USA: ASTM.

Aureli, F., Ciprotti, M., D'Amato, M., da Silva, E., Nisi, S., Passeri, D., Sorbo, A., Raggi, A., Rossi, M. and Cubadda, F. (2020). Determination of total silicon and SiO_2 particles using an ICP-MS based analytical platform for toxicokinetic studies of synthetic amorphous silica. *Nanomaterials*, 10(5): 888.

Baer, D.R. (2020). Guide to making XPS measurements on nanoparticles. *Journal of Vacuum Science & Technology A*, 38(3): 031201.

Baladin Group. (2022, May 30). *Thermal Characterization of Materials.* From Importance of Thermal Characteristics of Materials: https://balandingroup.ucr.edu/expertise_ThermalCharacterization.html.

Barabadi, H., Mojab, F., Vahidi, H., Marashi, B., Talank, N., Hosseini, O. and Saravanan, M. (2021, July). Green synthesis, characterization, antibacterial and biofilm inhibitory activity of silver nanoparticles compared to commercial silver nanoparticles. *Inorganic Chemistry Communications*, 129(July 2021): 108647.

Barreca, D., Gri, F., Gasparotto, A., Altantzis, T., Gombac, V., Fornasiero, P. and Maccato, C. (2018). Insights into the plasma-assisted fabrication and nanoscopic investigation of tailored MnO_2 nanomaterials. *Inorganic Chemistry*, 57(23): 14564–14573.

Bhagyaraj, S.M. and Oluwafemi, O.S. (2018). Nanotechnology: The science of the invisible. pp. 1–18. *In*: Bhagyaraj, S.M. (ed.). *Synthesis of Inorganic Nanomaterials - Advances and Key Technologies— A volume in Micro and Nano Technologies.* Woodhead Publishing.

BioSAXS. (2022, May 19). *BioSAXS.* From Small angle X-ray scattering: https://biosaxs.com/technique.html

Brongersma, H.H., Grehl, T., Schofield, E.R. and Smith, R.A. (2010). Analysis of the outer surface of platinum-gold catalysts by low-energy ion scattering. *Platinum Metals Review*, 54(2): 81.

Busby, Y., Agresti, A., Pescetelli, S., Di Carlo, A., Noel, C., Pireaux, J.-J. and Houssiau, L. (2018). Aging effects in interface-engineered perovskite solar cells with 2D nanomaterials: A depth profile analysis. *Materials Today Energy*, 9: 1–10.

Carneiro, Í., Viana, F., Vieira, M.F., Fernandes, J.V. and Simões, S. (2019). EBSD analysis of metal matrix nanocomposite microstructure produced by powder metallurgy. *Nanomaterials*, 9(6): 878–889.

Chen, Y., Smock, S.R., Flintgruber, A.H., Perras, F.A., Brutchey, R.L. and Rossini, A.J. (2020). Surface termination of CsPbBr3 perovskite quantum dots determined by solid-state NMR spectroscopy. *Journal of the American Chemical Society*, 142(13): 6117–6127.

Creative Biostructure. (2022, May 27). *Thermodynamics Analysis.* From Thermal Gravimetric Analysis (TGA): https://www.creative-biostructure.com/maghelix%E2%84%A2-thermal-gravimetric-analysis-tga-216.htm

Creative Proteomics. (2022, May 27). *Spectroscopy Technology.* From Differential Scanning Calorimetry Based Analysis Service: https://www.creative-proteomics.com/support/dsc-based-analysis-service.htm

Daw, J.E., Rempe, J.L., Knudson, D.L., Condie, K.G. and Crepau, J.C. (2008). *Viability of Pushrod Dilatometry Techniques for High Temperature In-Pile Measurements*. USA: US Department of Energy Office of Scientific and Technical Information.

de Oliveira Barbosa, W., Dourado Maia, R. and Dias Filho, J. (2020). Parametrization of magnetic nanoparticles - mathematical model using evolutionary algorithms. *Revista Mundi Engenharia Tecnologia e Gestão*, 5(6): 208-01–288-12.

Diculescu, V.C., Chiorcea-Paquim, A.-M., Corduneanu, O. and Olivera-Brett, A.M. (2007, February 14). Palladium nanoparticles and nanowires deposited electrochemically: AFM and electrochemical characterization. *Journal of Solid State Electrochemistry*, 11: 887–898.

Dutra, G., Canning, J., Padden, W., Martelli, C. and Dligatch, S. (2017). Large area optical mapping of surface contact angle. *Optics Express*, 25(18): 21127–21144.

Encyclopædia Britannica. (2022, May 7). *Encyclopædia Britannica*. From Scanning Electron Microscope: https://www.britannica.com/technology/scanning-electron-microscope#/media/1/526571/110970.

Fakhari, S., Jamzad, M. and Fard, H.K. (2019). Green synthesis of zinc oxide nanoparticles: A comparison. *Green Chemistry Letters and Reviews*, 12(1): 19–24.

Goyal, M. and Gupta, B. (2018). Shape, size and temperature dependency of thermal expansion, lattice parameter and bulk modulus in nanomaterials. *Pramana*, 90: 80.

Grimm Group. (2022, May 23). *Grimm Group*. From XPS and UPS Background: https://grimmgroup.net/research/xps/background/.

Hachtel, J.A., Lupini, A.R. and Idrobo, J.C. (2018). Exploring the capabilities of monochromated electron energy loss spectroscopy in the infrared regime. *Scientific Reports*, 8: 5637.

Hitachi. (2022, May 22). *Thermal Analysis*. From Principle of Dynamic Mechanical Analysis (DMA): https://www.hitachi-hightech.com/global/products/science/tech/ana/thermal/descriptions/dma.html.

Hu, Z. (2017). Characterization of materials, nanomaterials, and thin films by nanoindentation. pp. 165–239. *In:* Thomas, S., Thomas, R., Zachariah, A.K. and Mishra, R.K. (eds.). *Microscopy Methods in Nanomaterials Characterization*. Science Direct.

ISO. (2017). ISO 22412. *Particle Size Analysis – Dynamic Light Scattering (DLS)*. International Organization for Standardization.

Jasco. (2022, May 11). *Jasco*. From Principles of infrared spectroscopy (4) Advantages of FTIR spectroscopy: https://www.jasco-global.com/principle/principles-of-infrared-spectroscopy-4-advantages-of-ftir-spectroscopy/.

Jasrotia, R., Singh, V., Kumar, R. and Singh, M. (2020). Raman spectra of sol-gel auto-combustion synthesized Mg-Ag-Mn and Ba-Nd-Cd-In ferrite based nanomaterials. *Ceramics International*, 46(1): 618–621.

Jeol. (2022, May 19). *Nuclear Magnetic Resonance Spectrometer (NMR)*. From NMR basic knowledge: https://www.jeol.co.jp/en/products/nmr/basics.html.

Jeol. (2022, May 29). *Glossary of SEM Terms*. From Electron Backscatter Diffraction, EBSD: https://www.jeol.co.jp/en/words/semterms/search_result.html?keyword=electron%20backscatter%20diffraction%2C%20EBSD.

Konde, S., Ornik, J., Prume, J.A., Taiber, J. and Koch, M. (2020). Exploring the potential of photoluminescence spectroscopy in combination with Nile Red staining for microplastic detection. *Marine Pollution Bulletin*, 159(Octubre 2020): 111475.

Li, C., Cauwe, M., Yang, Y., Schaubroeck, D., Mader, L. and Op de Beeck, M. (2019). Ultra-long-term reliable encapsulation using an atomic layer deposited HfO2/Al2O3/HfO2 triple-interlayer for biomedical implants. *Coatings*, 9(9): 579.

Liu, C., Qiu, S., Du, P., Zhao, H. and Wang, L. (2018). An ionic liquid–graphene oxide hybrid nanomaterial: Synthesis and anticorrosive applications. *Nanoscale*, 10: 8115–8124.

Liu, X., Tang, B., Long, J., Zhang, W., Liu, X. and Mirza, Z. (2018). The development of MOFs-based nanomaterials in heterogeneous organocatalysis. *Science Bulletin*, 63(8): 502–524.

Lohse, S.E., Burrows, N.D., Scarabelli, L., Liz-Marzán, L.M. and Murphy, C.J. (2014). Anisotropic noble metal nanocrystal growth: The role of halides. *Chemistry of Materials*, 26(1): 34–43.

Luo, P., Roca, A., Tiede, K., Privett, K., Jiang, J., Pinkstone, J., Ma, G., Veinot, J. and Boxall Alisatair. (2018, February). Application of nanoparticle tracking analysis for characterising the fate of engineered nanoparticles in sediment-water systems. *Journal of Environmental Sciences*, 64(February 2018): 62–71.

Mahmoudi, F., Farhadi, S., Jarosova, M. and Sillannpää, M. (2020). Preparation of novel hybrid nanomaterials based on LaFeO3 and phosphotungstic acid as a highly efficient magnetic photocatalyst for the degradation of methylene blue dye solution. *Applied Organometallic Chemistry*, 34(12): e6011.

Mehrali, M., Latibari, S.T., Mehrali, M., Mahlia, T.I., Sageghinezhad, E. and Metselaar, H.C. (2014). Preparation of nitrogen-doped graphene/palmitic acid shape stabilized composite phase change material with remarkable thermal properties for thermal energy storage. *Applied Energy*, 135, 339–349.

Meretics. (2022, May 17). *Measure, Analyse, Control*. From BET Surface Area Analysis: https://www.meritics.com/technologies/bet-surface-area-analysis/.

Micromeretics. (2022, May 23). *Surface Area*. From How Specific Surface Area Is Important To Your Products Performance: https://www.micromeritics.com/particle-testing/analytical-testing/surface-area/.

Montana State University. (2022, May 19). *Nanotechnology in STEM*. From Secondary Ion Mass Spectrometer (SIMS): https://serc.carleton.edu/msu_nanotech/methods/SIMS.html.

Moore Analytical. (2022, May 18). *X-ray Diffraction (XRD)*. From X-ray Diffraction (XRD): https://www.mooreanalytical.com/x-ray-diffraction-xrd/.

Mousavihashemi, S., Lahtinen, K. and Kallio, T. (2022). *In-situ* dilatometry and impedance spectroscopy characterization of single walled carbon nanotubes blended LiNi0.6Mn0.2Co0.2O2 electrode with enhanced performance. *Electrochimica Acta*, 412(April 2022): 140093.

nanoComposix. (2022, May 29). *NanoComposix*. From Zeta Potential Measurements: https://nanocomposix.com/pages/zeta-potential-measurements.

NanoScience Instruments. (2022, May 31). *NanoScience Instruments*. From Optical Profilometry: https://www.nanoscience.com/techniques/optical-profilometry/.

Neumeier, J. (2015). *Photophysics of Graphene Quantum Dots*. Max Planck Institute of Quantum Optics.

Nideep, T.K., Ramya, M., Nampoori, V. and Kailasnath, M. (2020). The size dependent thermal diffusivity of water soluble CdTe quantum dots using dual beam thermal lens spectroscopy. *Physica E: Low-dimensional Systems and Nanostructures*, 116(February 2020): 113724.

NIST. (2022, May 8). *Material Measurement Laboratory*. From Nano-Measurements: Complete List of Protocols: https://www.nist.gov/mml/nano-measurements-complete-list-protocols.

Particle Technology Labs. (2022, DSC 28). *Testing*. From Differential Scanning Calorimetry (DSC): https://www.particletechlabs.com/analytical-testing/thermal-analyses/differential-scanning-calorimetry.

Perez-Potti, A., Lopez, H., Pelaz, B., Abdelmonem, A., Soliman, M.G., Schoen, I., Kelly, P.M., Dawson, K.A., Parak, W.J., Krpetic, Z. and Monopoli, M.P. (2021). In depth characterisation of the biomolecular coronas of polymer coated inorganic nanoparticles with differential centrifugal sedimentation. *Scientific Reports*, 11: 6443.

Peskersoy, C. and Culha, O. (2017). Comparative evaluation of mechanical properties of dental nanomaterials. *Journal of Nanomaterials*, 2017: 6171578.

Queffélec, C., Forato, F., Bujoli, B., Knight, D.A., Fonda, E. and Humbert, B. (2020). Investigation of copper oxidation states in plasmonic nanomaterials by XAS and Raman spectroscopy. *Physical Chemistry Chemical Physics*, 22: 2193–2199.

Rafique, M. (2015, March). Study of the Magnetoelectric Properties of Multiferroic Thin Films and Composites for Device Applications. Islamabad, Pakistan: COMSATS Institute of Information Technology Islamabad-Pakistan.

Raja, P.M. and Barron, A.R. (2020). *Physical Methods in Chemistry and Nano Science*. Rice University.

Raman Spectrocopy. (2022, May 24). From Stokes and Anti-Stokes: https://www.youtube.com/watch?v=SsIYDEma_cU.

Rattanawongwiboon, T., Soontaranon, S., Hemvichian, K., Lertsarawut, P., Laksee, S. and Picha, R. (2022, February). Study on particle size and size distribution of gold nanoparticles by TEM and SAXS. *Radiation Physics and Chemistry*, 191(February 2022): 109842.

Rikken, R.S., Nolte, R.J., Maan, J.C., van Hest, J.C., Wilson, D.A. and Christiansen, P.C. (2014). Manipulation of micro- and nanostructure motion with magnetic fields. *Soft Matter*, 2014(10): 1295–1308.

Saha, B. and Chakraborty, P. (2013). MCsn+-SIMS: An innovative approach for direct compositional analysis of materials without standards. *Energy Procedia*, 41(71): 80–109.

Sarto, G., Lopes, F., dos Santos, F.R., Parreira, P.S. and Almeida, L.C. (2019, September). Characterization of Cu2O/TiO2NTs nanomaterials using EDXRF, XRD and DRS for photocatalytic applications. *Applied Radiation and Isotopes*, 151(September 2019): 124–128.

Schwartzberg, A.M., Olson, T.Y., Talley, C.E. and Zhang, J.Z. (2006). Synthesis, characterization, and tunable optical properties of hollow gold nanospheres. *The Journal of Physical Chemistry BB*, 110: 19935–19944.

Sendra, M., Volland, M., Balbi, T., Fabbri, R., Yeste, M.P., Gatica, J.M., Canesi, L. and Blasco, J. (2018). Cytotoxicity of CeO2 nanoparticles using *in vitro* assay with Mytilus galloprovincialis hemocytes: Relevance of zeta potential, shape and biocorona formation. *Aquatic Toxicology*, 200(July 2018): 13–20.

Šetka, M., Calavia, R., Vojkuvka, L., Llobet, E., Drbohlavová, J. and Vallejos, S. (2019). Raman and XPS studies of ammonia sensitive polypyrrole nanorods and nanoparticles. *Scientific Reports*, 9: 8465.

Shibata, Y., Nomura, S., Kashiwaya, H., Kashiwaya, S., Ishiguro, R. and Takayanagi, H. (2015). Imaging of current density distributions with a Nb weak-link scanning nano-SQUID microscope. *Scientific Reports*, 5: 15097.

SIGRAY. (2022, May 28). *X-ray Absorption Spectroscopy (XAS)*. From X-ray Absorption Spectroscopy (XAS): https://sigray.com/x-ray-absorption-spectroscopy/#:~:text=X%2Dray%20absorption%20spectroscopy%20(XAS)%20is%20a%20chemical%20state,close%20to%20the%20absorption%-20edge.

Skrátek, M., Dvurecenskij, A., Kluknavsky, M., Barta, A., Balis, P., Micurová, A., Cigán, A., Eckstein-Andicsová, A., Manka, J. and Bernátová, I. (2020). Sensitive SQUID bio-magnetometry for determination and differentiation of biogenic iron and iron oxide nanoparticles in the biological samples. *Nanomaterials*, 10(10): 1993.

SMACgigWORLD. (2022, May 23). *Principle of UV - Vis Spectroscopy*. From Spectroscopy and UV-Vis Absorption Variant: https://www.smacgigworld.com/blog/principle-uv-vis-spectroscopy-.php.

Terzano, R., Denecke, M., Falkenberg, G., Miller, B.W., Paterson, D. and Jansenss, K. (2019). Recent advances in analysis of trace elements in environmental samples by X-ray based techniques. *Pure and Applied Chemistry*, 91(6): 1030–1060.

Thomson, T. (2014). Magnetic properties of metallic thin films. pp. 454–546. *In*: Barmak, K. and Coffey, K. (eds.). *Metallic Films for Electronic, Optical and Magnetic Applications: Structure, Processing and Properties*. Woodhead Publishing.

Tian, M., Dyck, O., Ge, J. and Duscher, G. (2019). Measuring the areal density of nanomaterials by electron energy-loss spectroscopy. *Ultramicroscopy*, 196(January 2019): 154–160.

Tofwerk. (2022, May 13). *Tofwerk*. From What is Single Particle ICP-MS? https://www.tofwerk.com/single-particle-icp-ms/.

TWI. (2022, May 16). *Technical Knowledge*. From What is X-Ray Diffraction Analysis (XRD) and how does it work? https://www.twi-global.com/technical-knowledge/faqs/x-ray-diffraction.

Vallejo, J.P., Żyla, G., Fernández-Seara, J. and Lugo, L. (2019). Influence of six carbon-based nanomaterials on the rheological properties of nanofluids. *Nanomaterials*, 9(2): 146–164.

Vetrivel, S., Sri Abirami Saraswathi, M., Rana, D. and Nagendran, A. (2018). Fabrication of cellulose acetate nanocomposite membranes using 2D layered nanomaterials for macromolecular separation. *International Journal of Biological Macromolecules*, 107(Part B): 1607–1612.

Wang, F. and Liu, X. (n.d.). Upconversion multicolor fine-tuning: Visible to near-infrared emission from lanthanide-doped NaYF4 nanoparticles. *Journal of American Chemistry Society*, 130(17): 5642–5643.

Wang, S., Ye, B., An, C., Wang, J. and Li, Q. (2018). Synergistic effects between Cu metal–organic framework (Cu-MOF) and carbon nanomaterials for the catalyzation of the thermal decomposition of ammonium perchlorate (AP). *Journal of Materials Science*, 54: 4928–4941.

Wilschefski, S.C. and Baxter, M.R. (2019). Inductively coupled plasma mass spectrometry: Introduction to analytical aspects. *The Clinical Biochemist Reviews*, 40(3): 115–133.

Woodhead Publishing. (2017). Rheological characteristics of nanomaterials and nanocomposites. pp. 327–350. *In*: Abraham, J., Shakira, T., Mishra, R.K. and Thomas, S. (eds.). *Micro and Nano Fibrillar Composites (MFCs and NFCs) from Polymer Blends*.

Wu, Q., Miao, W.-S., Zhang, Y.-D., Gao, H.-J. and Hui, D. (2020). Mechanical properties of nanomaterials: A review. *Nanotechnology Reviews*, 9(1): 259–273.

Yadav, A.K., Banerjee, S., Kumar, R., Kar, K.K., Ramkumar, J. and Dasgupta, K. (2018). Mechanical analysis of nickel particle-coated carbon fiber-reinforced epoxy composites for advanced structural applications. *ACS Applied Nano Materials*, 1(8): 4332–4339.

Zainabb, S.R. (2015). *Electron Backscatter Diffraction of Gold Nanoparticles*. [Master's Thesis, McMaster University]. McMaster University. https://macsphere.mcmaster.ca/bitstream/11375/18237/2/SRZainab%20Thesis%20-%20Electron%20Backscatter%20Diffraction%20of%20Gold%20Nanoparticles.pdf.

Zhang, D., Zheng, Y., Yuan, G., Qian, G., Zhang, H., You, Z. and Li, P. (2022). Chemical characteristics analyze of SBS-modified bitumen containing composite nanomaterials after aging by FTIR and GPC. *Construction and Building Materials*, 324: 126522.

Zhang, H., Wang, C. and Zhou, G. (2020). Ultra-microtome for the preparation of TEM specimens from battery cathodes. *Microscopy and Microanalysis*, 26(5): 1–11.

Chapter 4

Nanomaterials under High-temperature Conditions

Paulina L. Ríos,[1] Paulo Preuss,[2] Juan Luis Arroyo,[3] Paula Povea[3] and María Belén Camarada[2,4,]*

1. Introduction

The use of nanoparticles in different industrial processes like catalysis is well known. The upper surface-to-volume ratio compared to microparticles increases their activity, and the quantum confinement can alter the chemical characteristics of sufficiently small nanoparticles (Narayanan and El-Sayed 2005). The melting point of metal nanoparticles decreases with smaller sizes, and one common problem for nanoparticles is their lack of stability at critical conditions like extremely high temperatures. For many processes, such as catalytic combustion, steam reforming, and industrial production, reaction temperatures are typically over 600°C, and the thermal stability of the material becomes crucial (Zarur and Ying 2000). On the other hand, exploration of outer space requires the design of new nanomaterials resistant to high temperature and pressure. This chapter reviews some typical applications of nanomaterials in applications that need high temperatures, like industrial tools, sensors, and energetic materials.

[1] Centro de Nanotecnología Aplicada, Facultad de Ciencias, Ingeniería y Tecnología, Universidad Mayor, Camino La Pirámide 5750, Huechuraba, Santiago, RM.

[2] Laboratorio de Materiales Funcionales, Departamento de Química Inorgánica, Facultad de Química y de Farmacia, Pontificia Universidad Católica de Chile, Santiago, 7820436, Chile.

[3] Laboratorio de Materiales Energéticos, Instituto de Investigaciones y Control del Ejército de Chile (IDIC), Av. Pedro Montt 2136, 8370899, Santiago, Chile.

[4] Centro Investigación en Nanotecnología y Materiales Avanzados, CIEN-UC, Pontificia Universidad Católica de Chile, Santiago, Chile.

* Corresponding author: mbcamara@uc.cl

2. Nanostructured Materials for High-temperature Applications

2.1 Heat-resistant Nanocomposites as Material for Tools

The search for innovative materials is continuously developed in many applications like cutting, forming, and casting tools. The most important properties for devices are hardness, toughness, and chemical stability. These properties are relevant in machining operations on hard and tough materials, especially in extreme conditions like interrupted cutting, high-speed cutting, and dry cutting operations. In particular, dry machining is of enhanced research interest because of its environmental benefits. However, the high heat generation during the process may weaken the desirable tool properties leading to failure. Up to date, one of the available solutions with the best results is the modification of surfaces through the deposition of protective coatings to improve tools and components' lifetime and performances (Gekonde and Subramanian 2002, Mitterer et al. 2000, Sandstrom and Hodowany 1998). Therefore, the oxidation resistance of protective coatings at elevated temperatures becomes a critical focus of interest. Due to the high hot hardness and toughness, mixed alumina is suitable for hard machining purposes (Arunachalam et al. 2004). Ceramics can be mixed with alumina, but vibrations and forces may affect the final behavior (Aslantas et al. 2012).

In the area of nanotechnology, most efforts have been devoted to the development of super hard nanocomposite thin films like nanocrystalline transition metal nitrides dispersed in an amorphous silicon nitride (Veprek 1999). The size reduction down to the nanometer scale leads to significant changes in physical and mechanical properties, leading to major improvements. During the last years, coated ceramic thin films have proved their efficiency and have received necessary attention due to their superior chemical stability at temperatures up to 800°C. The binary nitride system, TiN, is the most widely applied as protective coatings. The production of TiN with physical vapor deposition (PVD) started in the late 70s, based on electron beam ion plating technology. Since then, PVD coatings and their application technology have represented a significant area in the scope of nanotechnology and allowed the rapid development of nanocomposites (Rodríguez-Barrero et al. 2016). One of the main disadvantages of TiN is the fast oxidation at temperatures above 500°C, which reduces the range of its application. The improvement has been achieved by developing ternary and quaternary nitride systems. During the late 90s, the addition of aluminum to the TiN base composition to give TiAlN provided not only a higher hardness but also a remarkably enhanced behavior at extreme conditions, thermal stability, and oxidation resistance at high temperatures, with application in high-speed machining tools (Chen et al. 2012, Du et al. 2003, PalDey and Deevi 2003). Compared to TiN, TiC, and CrN, TiAlN coatings have better oxidation resistance, higher hardness, and improved chemical stability at higher temperatures due to forming a protective Al_2O_3 layer at high temperatures (Hörling et al. 2005). However, the mechanical properties and oxidation resistance decrease significantly at a high-temperature environment due to structural change from cubic (B1) to hexagonal (B4) phase (Fuentes et al. 2010). The evolution of TiAlN to AlTiN coatings, with higher aluminum content and, therefore, better thermal and corrosion resistance, leads to a further increase in the lifetime of tools under high-speed cutting (Rodríguez-Barrero et al. 2016). AlTiN coatings have been offered as a tool coating since 2011 (Bobzin 2017). Lately, the addition of silicon (TiAlN–TiSiN) forming a silicon nitride binder surrounding the TiAlN crystals ensured a fine nanostructure up to 1200°C. In this way, the usual hardness loss at high temperatures was minimized, making them extremely suitable for adverse cutting applications (Veprek et al. 2004, Wang et al. 2010).

Also, TiAlSiN based nanocomposite coatings are of high interest due to their chemical stability at elevated temperatures (Holubář et al. 1999). At lower Si content, the coating accounts for higher coating adhesion to the substrate, whereas with medium Si content, the microhardness of this coating increases, but the coating/substrate adhesion decreases (Yu et al. 2009). This problem can be resolved by adding some other phases like TiAlN on the tool substrate before the deposition of TiAlSiN coating (Sui et al. 2016). The excellent mechanical properties and thermal stability of TiAlSiN coatings deposited by an industrial vacuum arc evaporation were reported by Veprek et al. (Veprek et al. 2004), while He and coworkers produced TiAlSiN coatings with different Al and Si contents using hybrid PVD technology. The friction and wear of TiAlSiN coatings show a strong dependence on stoichiometry and test temperature. The formation of $SiO_2 \cdot nH_2O$ contributes to decreasing the wear rate and friction coefficient at room temperature. The rutile TiO_2 phase produced at 800°C plays the role of lubrication in the process of sliding (He et al. 2016).

Recently, oxygen was added to nitride coatings to increase oxidation and adhesion resistance against the workpiece material (Bobzin et al. 2017). (Ti,Al,Cr,Si) ON coatings deposited on tungsten carbide substrates show high phase stability up to a temperature of T = 1200°C and high oxidation resistance up to T = 900°C (Figure 1). Overall, the coatings show high potential for application on cutting tools for high-performance cutting (Bobzin et al. 2019).

In the 2000s, AlCrN coatings expanded the capabilities of TiAlN coatings. The hardness of AlCrN coating is similar to TiAlN (Chim et al. 2009), but has a higher oxidation resistance (up to 1200°C), related to the growth of a stable $(Al,Cr)_2O_3$ oxide during the cutting process instead of the TiO_2 and Al_2O_3 oxides growth in TiAlN coatings. When both high-temperature and oxidation resistance are required, silicon coating containing AlCrSiN is preferred. Under long-time heat treatment at 1000°C, this nanocomposite matrix can provide a stable phase, and excellent oxidation resistance since the dense structure can effectively hinder the inward and outward diffusion of oxygen and chromium, respectively (Chen et al. 2011).

From a material point of view, TiAlN coatings with different alloying elements have endless opportunities. In this sense, high-entropy alloys (HEA) are defined as materials containing at least five principal elements with concentrations between 5 at.% and 35 at.% (Lai et al. 2007), where high entropy of mixing can stabilize the formation of a single-phase solid solution and prevent the formation of the intermetallic compound during solidification. Lattice distortions caused by different sizes of atoms decrease the coefficient of diffusion of the atoms, reducing the growth of crystallites as a result. HEA nitride coatings (TiHfZrVNb)N with a single cubic (NaCl-type) nitride phase revealed oxidation-resistant to 600°C (Pogrebnjak

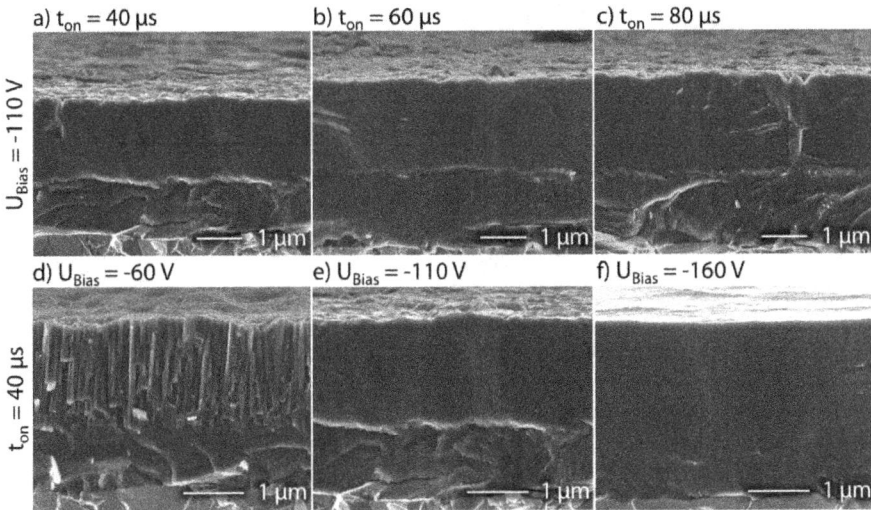

Figure 1. SEM cross-sections of (Ti, Al, Cr, Si)ON (a–c) and (Ti, Al, Cr, Si)N (d–f) after heat treatment for t = 0,5 h in vacuum at T = 1100°C in contact with a steel counterpart AISI M2 under a contact pressure of p = 210 kPa, T = 1200°C and T = 1300°C without a counterpart. Figure extracted from reference (Bobzin et al. 2019).

et al. 2014). (AlCrTiSi)N, one of the high-performance HEA nitride coating systems, shows an excellent hardness and wear resistance combination. Its application at high-speed and dry cutting conditions, with temperatures above 700°C, reveals a higher cutting performance when compared with TiSiN, TiAlSiN, AlCrN, and AlCrTiN coated cutting tools (Chen et al. 2018). Recently, carbon nanotubes (CNTs) reinforcing HEA composites were fabricated on a TA2 titanium alloy with laser melting deposition of FeCoCrAlCu-(nano-SiB6)-(Ni/Ag-coated CNTs) mixed powders, noting that large quantities of the ultrafine nanocrystals were formed (Figure 2). These nanocrystals could permeate into the originally ordered atom arrangement due to the high diffusion rates and their ultrafine microstructure, favoring a compact structure. The CNTs/ultrafine nanocrystals composites showed better corrosion and high-temperature oxidation resistance (T = 800°C) than TA2 alloy. This study provides fundamental experimental information to upgrade the quality of industrial light alloys (Li et al. 2020).

Furthermore, literature shows that the structure and deposition layers play a significant role in defining the performance of coatings. Nano-multilayered films are of increasing interest due to their superior mechanical and lubrication properties in advanced tribological applications where single-layer films are insufficient (Çalışkan et al. 2013, Sui et al. 2016). TiAlN/CrAlN nano-multilayer coatings with different bilayer periods have been assembled to improve the properties of TiAlN and CrAlN single-layer coatings. Oxidation resistance is increased with a higher ratio of CrAlN. The multilayer structure combines the advantages of thermal stability for TiAlN and oxidation resistance for CrAlN, and increased hardness and adhesive strength (Li et al. 2013). Nano-multilayered TiAlCrSiN films, consisting of alternating crystalline TiCrN and AlSiN nanolayers, were deposited on a steel substrate by the cathodic

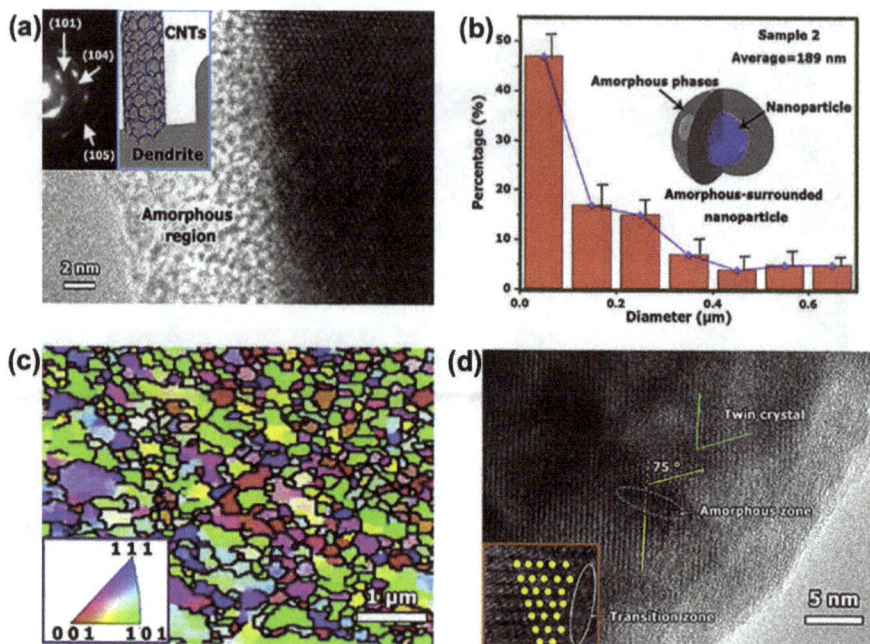

Figure 2. (a) High-resolution transmission electron microscopy (HRTEM) and the corresponding SAED pattern of FeCoCrAlCu-(nano-SiB6)-(Ni/Ag-coated CNTs) sample, showing CNTs attached to dendrites, (b) size distribution of the coating in the sample, (c) EBSD and (d) HRTEM patterns in the coating. Figure adapted from reference (Li et al. 2020).

arc plasma process. TiAlCrSiN films oxidized slower than the TiN films and faster than the CrN or CrAlSiN films. Despite long oxidation at 900°C and the escape of nitrogen from the film, the oxidation resistance of TiAlCrSiN film was good due to the formation of Cr_2O_3, α-Al_2O_3, and SiO_2 (Nguyen et al. 2009). The effect of multilayer architecture on the structure and thermal stability of TiSiN/TiN nanocomposite coating has also been investigated. The TiSiN coating with nanocrystalline grains encapsulated by the amorphous matrix (TiN of 6.6 nm and 3.7 nm TiSiN layers) maintains the hardness upon annealing to 1100°C, where the hardness of TiSiN coating decreases due to the grain coarsening and the thermal decomposition (Xu et al. 2017). More recently, multilayered TiSiN/TiN(Ag) coatings have been explored as self-lubricant films that combine the intrinsic properties of high oxidation resistance with lubricious elements deposited as a monolayer or multilayer configuration. The coatings showed excellent thermal stability near 700°C and revealed a good control of the Ag diffusion towards the surface in a protected atmosphere (AL-Rjoub et al. 2020).

Considering the information mentioned above, significant efforts have been made and must continue to obtain materials that lead to the highest possible productivity levels. Manufacturers are still working on novel materials with superior capacities of strength and cost that can successfully replace today's best benchmark options. The future points of nanocomposites for machining tools are hybrid structures capable of enhancing performance. More studies in dislocation, deformation, and mathematical modeling are needed to develop new materials with optimal properties.

2.2 *Materials for Friction Assemblies*

Energy waste by heat generation in machining processes such as drilling, milling, turning, and grinding, among others, due to friction between the tool and the workpiece can lead to substandard efficiency of equipment and materials, like tool chipping, to a decrease of the tool wear resistance, low product quality, constant repair or change of pieces, and other increasing economic losses. For this reason, it is indispensable to control the machining temperatures, either by incrementing the hardness of materials or by reducing friction using a proper lubricant.

Traditional lubricating fluids are often commercial synthetic oils and have been conventionally used to flood the machining zone, thus reducing friction and acting as coolants. Nevertheless, its use has various negative consequences such as increased costs due to intensive use, environmental pollution during use and disposal, and safety hazards for operators. All of this, plus an increasing restriction of environmental regulations, encouraged research in the use of dry machining, minimum quantity lubrication (MQL), solid lubricants, and new biodegradable lubricants, improved tribological properties of lubricants by adding nanoparticles.

Dry machining does not use lubricating fluids and operates with tools made of different materials. New materials coated tools use dry machining, but they can reach high machining temperatures. MQL technique conditions use the lubricating fluid as an aerosol or mist in the machining environment, maintaining the lubricant quantity at a minimum. This technique effectively reduces machining temperature, but it can still present safety risks for operators, such as toxic mist. The use of biodegradable lubricants includes vegetable oils that can reduce environmental pollution and safety hazards, but do not have good tribological qualities as synthetic lubricants. Thus, research in the use of nanoparticles can result in a new nanolubricant development.

Nanolubricants or nanofluids are composed of nanoparticles like metal, metal oxide, metal sulfides, carbon as nanotubes, and other substances, of a size smaller than 100 nm, a base fluid such as water, vegetable or mineral oil, and may include additives or surfactants. Nanoparticles improve the tribological properties of the fluids and thermochemical characteristics. As stated by Lee and coworkers in nanolubricants, nanoparticles function as friction modifiers. Their possible action mechanism can be the rolling effect, the formation of a protective film, the mending, or the polishing effect (Lee et al. 2009). This section reviews some examples of nanoparticles used in different fluids as nanolubricants under extreme high-temperature conditions.

Metal nanoparticles like Cu had challenged researchers due to a poor interaction between the nanoparticle and the base fluid, leading to stabilization problems for sedimentation in long term storage; Ali and Xianjun (Ali and Xianjun 2020a) showed that Cu nanoparticles (10–20 nm), modified by the ionic liquid bis(2-ethylhexyl) phosphate (HDEHP), provides superior dispersion in base fluid polyalphaolefin-6 oil (PAO6), demonstrating that the nanolubricant behaved like typical non-Newtonian fluid and that the dynamic viscosity increased as the Cu concentration increased in the base fluid mixture (0.01 to 0.1 wt% Cu in 2 wt% HDEHP in PAO6). Thermogravimetric analysis (TGA) curves showed that the decomposition temperatures of the nanolubricants were higher than 280°C, presenting excellent thermal stability compared with the oil mixture without Cu (225°C). Compared to unmodified Cu nanoparticles, the nanolubricant did not present sedimentation for up to 60 days. It suggested an increase in the attractive van der Waals forces between the C due to positive interactions between its surface and the HDEHP, which modified it, improving its dispersity in nonpolar solvents. The probably high cost of this kind of nanoparticles makes their application somehow scarce.

Metal oxide nanoparticles such as Al_2O_3, ZnO, NiO, CaO, and mixtures like Al_2O_3/TiO_2 are widely studied since they are probably easier to find and at a lower cost than copper. Shen and coworkers used Al_2O_3 (40 nm) in a water-based synthetic fluid (Cimtech 500, 5 vol%) at 1, 2.5, and 4 vol% concentrations in a grinding wheel at wet, dry, and MQL conditions assessing tribological characteristics (Shen et al. 2008). The authors found that the dry grinding reached 595°C, and the best nanofluid reached 455°C (4 vol%, MQL), but the lowest temperature was achieved by flood cooling with Cimtech 500, 5 vol% (110°C). This is probably because the amount of nanofluids applied in MQL is too small to make any difference even though they have better heat transfer and thermal conductivity, showing that operating conditions are as important as nanolubricant tribological characteristics. Mao and coworkers also used Al_2O_3 nanoparticles (40 nm) but in deionized water at 1.2 wt% in MQL conditions, grinding a hardened AISI 52100 steel workpiece (Mao et al. 2014). The temperature in dry grinding reached up to approx. 550°C. With the use of a nanolubricant, the temperature decreased approx. 100°C, and was 40°C lower than using pure water, showing that water-base fluid nanolubricants can also achieve efficient lubrication in the grinding zone.

Other metal oxides like milled garnet mineral (45 nm), mostly composed of Si, Fe, and Al oxides, containing Ti, Mg, Ca, and Mn oxides, and others, dispersed in SN500 commercial lubricant oil have been explored. Melting temperatures for the SN500 lubricant oil, 0.25, and 0.75 wt% nanolubricants were 255°C, 410°C, and 425°C. This shows that nanolubricants considerably increased the ability to resist mass loss and the melting/decomposition temperatures (Maheswaran and Sunil 2016). Bhaumik and coworkers used ZnO (35–45 nm) dispersed in castor oil. The tribo-tests showed that both the anti-wear and extreme pressure properties of the castor oil were enhanced with ZnO up to 0.1 wt%. The determination of the flash and fire point showed that all nanolubricants decreased flash and fire point compared to castor oil by 10°C and 5°C from 270°C and 275°C, respectively (Bhaumik et al. 2018).

Sunil et al. assessed the thermal stability of NiO (~ 73 nm diameter) in SN500 commercial oil. Differential scanning calorimetry (DSC) showed that the thermal degradation temperature of the SN500 oil appeared at 49°C, while with 0.1, 0.3, and 0.5 wt.% nanolubricants were 210°C, 222°C, and 247°C, respectively (Sunil et al. 2019a). Other authors also used dispersed NiO (~ 55 nm) in SN500 (Ramachandran et al. 2020). At 200°C, SN500 oil loses 9.5% of the mass, whereas 0.1 and 0.3 wt% nanolubricant do not lose any mass at this temperature range. The thermostability of the base fluid enhances up to 0.3 wt%, and then, it decreases due to the agglomeration of dispersed nanoparticles. Sunil and coworkers used CaO (~ 38 nm) in rice bran oil. The thermal degradation of rice bran oil reached 388°C, and with 0.25 wt% rise to 414°C, showing better thermal stability (Sunil et al. 2019b).

Ali and coworkers used Al_2O_3/TiO_2 (10 nm and 8–12 nm) in a mixture of oleic acid in commercial engine oil, Castrol EDGE professional A5 5W-30 at final 0.1 wt% (0.05 wt% Al_2O_3, 0.05 wt% TiO_2, 1.9 wt% oleic acids) (Ali and Xianjun 2020b). The maximum decomposition of nanolubricant was 399°C, compared to the engine oil at 352.2°C, improving anti-decomposition properties. The use in an engine test showed that the thermal brake efficiency improved by 3.9–8.6% when using nanolubricants compared to the engine oil. This is related to the improved heat transfer capacity and thermal stability of nanolubricants. These metal oxide nanoparticles have a wider range of compatibility with base fluids and can be used without additives or surfactants. The authors also indicated that nanolubricants have an optimum concentration and above that point, nanoparticles tend to aggregate, decreasing tribological features.

Another type of nanoparticle is MoS_2 in different forms, like nanoballs or nanosheets. Kalita and coworkers used MoS_2 (< 100 nm) enhanced with triglycerides and phospholipids in paraffin oil to assess a grinding process in MQL and flood conditions. Among other characteristics, paraffin oil's peak grinding surface temperatures with MQL yielded 288°C, and the 2 and 8 wt% nanolubricants reduced peak temperature to 175°C and 160°C, respectively (Kalita et al. 2012).

Also, there are mixed nanoparticles in solid nanolubricants; Hua and coworkers used powder MoS_2 (40 nm), polyimide, and carbon nanotubes (CNTs, 3.5–12.9 nm), among others, to study the high-temperature friction and wear performance of micro dimple sliding contact surface of high temperature bearing steel. Solid nanolubricants mixtures of 80 wt% MoS_2, 20 wt% polyimide, and 6 wt% CNTs were smeared evenly on the contact surface. Results showed that MoS_2-PI solid nanolubricant present low friction coefficient in the temperature range from room temperature to 400°C, but MoS_2-polyimide-CNTs solid nanolubricant had lower friction coefficients in the same temperature range, reaching the lowest at 200°C. Due to the extremely high strength and extremely large toughness of CNTs, the strength and toughness of the composite solid lubrication film are improved. Besides, the good lubrication performance of the CNTs provides the benefits in improving the lubricating properties, showing that solid lubrication is also an alternative in nanolubrication (Hua et al. 2016).

Carbon nanoparticles such as graphite, nano-diamond, multi-walled carbon nanotubes (MWCNTs), and carbon nanotubes (CNTs) are also commonly researched; Srikiran and coworkers used graphite in SAE 40 oil to machining AISI 1040 steel under MQL conditions. The authors found that the optimum nanolubricant concentration was 0.5 wt% showing better tribological results. To assess the effect of the particle size, dry and 0.5 wt% nanolubricants with particles sizes of 5 to 90 nm were tested under different depths of cut. The dry condition temperature was approx. 140°C, while the best results were achieved with 70–90 nm nanolubricant. The temperature increased as particle size decreased, which emphasizes the decrease in lubrication performance in its nano-dimensions. This can be attributed to increased frictional resistance due to the aggregation of more nano-sized graphite particles at the tool–chip interface as the lubricant size decreases (Srikiran et al. 2013). Amrita and coworkers used graphite ($<$ 80 nm) in commercial oil Servo Cut S at 0.3 wt% in MQL conditions in turning AISI1040 steel, measuring cutting temperature, among others. The nanolubricant in MQL conditions reduced the temperature compared to flood machining because of the effective penetration at the interface regions due to aerosolization and improved thermal conductivity. Dry cutting reached approx. 200°C, while nanolubricant decreased the temperature to approx. 40°C (Amrita et al. 2014).

Sun and coworkers used surface-modified nano-diamond (50 nm) by dimer acid ester or oleic ester in colza oil at 0.5 wt% to assess high-temperature tribological performance from 100°C to 500°C. Results demonstrated no synergism between oleic acid and nano-diamond at low temperature. This may be caused by the inhibiting effect of the nano-diamond for the formation of the compact friction reactive film on the friction surface due to the strong adsorption of oleic acid. The friction coefficients of the modified nano-diamond by dimer ester are lower than those of nano-diamond and the modified nano-diamond by oleic ester when the temperature is lower than 200°C. For the high-temperature tribological property, the nano-diamond modified dimer acid ester is far superior to nano-diamond and the modified nano-diamond with oleic acid (Sun et al. 2013).

Pourpasha and coworkers used MWCNTs (5–16.1 nm) in a mixture with turbine oil, using Triton x100 as a surfactant to assess thermophysical attributes including viscosity, and flashpoint, among others. As the temperature increased from 30 to 100°C, the viscosity of nanofluids and pure oil declined by an average of ~ 85%. Increased temperature and molecular energy decrease intermolecular forces, leading to an increase in intermolecular distance. Then, increased temperature diminishes the viscosity of nanofluids. The flashpoint temperature increased, showing the maximum increment of 10°C compared to turbine oil (225°C), between 0.3–0.4 wt%. In general, raising the lubricant's resistance to ignition can be attributed to the increased thermal conductivity of the lubricant because of an increasing concentration of nanoparticles, and the addition of nanoparticles causes a delay in the vaporization of ignitable vapor (Pourpasha et al. 2020). Sharmin and coworkers used CNTs ($<$ 30 nm) stabilized with 0.6 vol% Sodium Dodecyl Sulfate (SDS) in deionized water using 10 vol% of Aquatex 3180 as a comparison to milling operation on 42CrMo4 hardened steel material. The testing on machining temperature was performed without any fluid, commercial fluid, and 0.3 m/v% nanofluid, selected for its higher stability, good thermal conductivity, and lower viscosity. The result showed a maximum 29% reduction in cutting temperature; dry machining reached 224°C after 9 minutes of operation, while commercial fluid lowered the temperature to 190°C, and the nanofluid to 159°C. The higher thermal conductivity of the nanofluid carried away the heat from the contact area of the tool and work, which less end the cutting temperature (Sharmin et al. 2020).

Other nanoparticles like boric acid and polyaniline nanotubes (PANI NTs) were also researched. A group used nanoboric acid (50 nm) suspended in SAE-40 or coconut oil. The thermal conductivity increased slightly with increasing concentration, contrary to the specific heat, while in both cases, higher values belonged to the coconut oil-based nanolubricants. Furthermore, the heat transfer coefficient of the coconut oil-based lubricants is greater than the SAE-40 oil-based. The heat transfer coefficient increased slightly with increasing concentration at a specific cutting speed. All coconut oil-based nanolubricant performed at the lowest cutting temperatures, reaching their optimum at approx. 163°C with a concentration of 0.5 wt%, while SEA-40 oil-based reached approx. 195°C at the same concentration (Vamsi Krishna et al. 2010). On the other hand, Sofiah et al. used PANI NTs (202–575 nm) dispersed in palm oil. Thermal degradation results reveal that palm oil begins at ~ 270°C, and a sudden degradation starts at 340°C and ends at 460°C, with a total degradation. The increasing concentrations of nanolubricants show a shift in the degradation curve, demonstrating that the nanolubricants can withstand high temperatures. 0.5 wt% nanolubricants demonstrate the most astounding movement in degradation temperature compared with base oil. The thermal conductivity measurement proved that the 0.5 wt% nanolubricant achieved the highest thermal conductivity of 0.4301 $Wm^{-1}K^{-1}$ at 80°C with an enhancement percentage of 25.76% compared to base oil at a similar temperature (Sofiah et al. 2020).

The type of nanoparticle may decide the dispersibility into the fluid since superficial charges may aggregate the particles, and the use of an additive or surfactant may change thermophysical properties. Likewise, the shape and size of the nanoparticles and the concentration are important in the development of a nanolubricant, since a lower concentration may not reach the lubricating effect, and too small nanoparticles at too high a concentration may aggregate, decreasing the lubricating effect. Also, the same nanoparticle may not behave the same, dispersed in different base fluids; so, to assess water, vegetable, or mineral oil accordingly, the nanoparticle type is also a consideration.

2.3 Sensors

Sensors are crucial for the correct functioning of equipment and monitoring different analytes in fields, such as medical detection, industrial production, and control of contaminants (Rodriguez et al. 2021, Ryu et al. 2021, Shao et al. 2020, Zhou et al. 2020). One important field is the detection of environmentally hazardous gases that are usually produced in high-temperature environments, such as NO_2, NO, N_2O, CO_2, H_2S, CO, NH_3, CH_4 and SO_2. Different kinds of gas sensors have been tested during the last decades based on special sensing materials and various transduction platforms. The main classes of materials include metal-oxide semiconductors, conducting polymers, conducting polymer composites, and metal-oxide/polymer composites, among others (Wetchakun et al. 2011). The gas sensors are based on various sensing mechanisms, such as optical, electrochemical, micro-electro-mechanical systems (MEMS), acoustic, surface acoustic wave, chemiresistive, etc. However, most of the reported devices work at moderate operating temperatures (under 250°C) (Ghosh et al. 2017, Maity et al. 2015), and there is an increasing interest in applications related to combustion systems, military, or nuclear sensor development that require sensing of gases at higher temperatures. In this sense, nanomaterials offer interesting solutions to be tested.

Up to date, only a few advances have been reported in the design of new sensors working at high temperatures using MEMS gas sensors. The number of prototypes has been increasing. Still, most of the reports need moderate temperatures to operate. In 2007, Meier and coworkers reported coupling nanowire chemiresistors with MEMS gas sensing platform based on a single nanowire SnO_2 obtained by vapor-solid growth method with the best response at 260°C (Meier et al. 2007). In 2008, Bhattacharyya worked on a nanocrystalline ZnO-based sensor using micromachined silicon substrate to detect methane. The best response was obtained at 250°C for 1% of the gas with a fast response of 8 s and a recovery time of 35 s (Bhattacharyya et al. 2008). Later, in 2016 Behera and Chandra published a sensor based on ZnO–CuO nanoflakes deposited on oxidized Si substrate to detect acetone, NH_3, ethanol, H_2S, and NO_2 at temperatures between 100 and 350°C (Behera and Chandra 2016). With reproducible and stable performance, a better identification was obtained for acetone at 300°C for 10 ppm. Compared with the other analyzed gases, acetone has a high dipolar moment and, thus, is more likely to be absorbed by the sensing material (ZnO and CuO), resulting in an enhanced response. A few years later, Kim and coworker tested nano-sized particles to detect different gases: $Pd–SnO_2$ for CO, In_2O_3 for NO_x, and $Ru–WO_3$ for NH_3. All MEMS gas sensor arrays showed good properties for their target gases with a temperature range of 270–300°C (Kim and Kim 2013).

In the detection of CO and CO_2, in 2006, a high-temperature CO gas sensor using Au nanoparticles embedded in an yttria-stabilized zirconia matrix was reported. The nanocomposite films had a surface plasmon resonance absorption band around 600 nm with a reversible blue shift and band narrowing upon exposure to CO in air ambient at 500°C. CO gas concentration varied in the range between 0.1 and 1 vol% showing a linear increment of the plasmonic band. The device showed no measurable signal change upon exposure to CO at temperatures below ~ 400°C, confirming the crucial role of temperature on the CO sensitivity (Sirinakis et al. 2006). In 2010, Lim and coworkers reported the synthesis of mesoporous In_2O_3 nanofibers with high surface area by the calcination of electrospun indium acetate/polyvinyl alcohol composites. The nanofibers had diameters in the range of 150–200 nm and consisted of cubic indium oxide nanocrystals with a primary particle size of 10–20 nm. The response of mesoporous In_2O_3 nanofibers to CO in the air was strongly affected by the surface area of the material, achieving a response of 500–100 ppm at 300°C. Gallium oxide (Ga_2O_3), often recognized as one of the widest-bandgap semiconductors, has shown remarkable potential as a gas receptor because of its chemical and thermal stability and excellent electronic properties (Afzal 2019). β-Ga_2O_3 nanorods have also been studied for the sensing of CO. Perovskite nanoparticle-sensitized Ga_2O_3 nanorod arrays were synthesized via a hydrothermal method, achieving 20–100 ppm detection at 500°C (Lin et al. 2016). A Ga_2O_3–In_2O_3 nanocomposite obtained by a co-precipitation method was applied as ethanol, CO, and CH_4 sensor. The responses and selectivity depended on the gallium content and operating and calcination temperatures. At 300°C, 10 wt% Ga_2O_3–In_2O_3 and 25 wt%

Ga_2O_3–In_2O_3 sensors calcined at 500°C showed maximum response to ethanol and CO, respectively, while the highest response to methane was observed at 400°C with 65 wt% Ga_2O_3. In addition, 25 wt% Ga_2O_3–In_2O_3 calcined at 850°C was selective to methane at 400°C (Bagheri et al. 2015). Lanthanum has also been tested as a gas sensor. In 2016, Jeong and coworkers reported 50% La-loaded ZnO nano-powder as a highly sensitive CO_2 chemi-resistive sensing system with a maximum response to 5000 ppm (65%) CO_2 at an operating temperature of 400°C (Jeong et al. 2016).

Sensors for NO_2 detection have been the most widely studied because NO_2 is considered the most dangerous gas. In 2009, Liu and coworkers reported the synthesis of WO_3 microspheres composed of orthorhombic $WO_3 \cdot xH_2O$ nanorods with a diameter less than 100 nm, which lost water gradually during annealing and changed to monoclinic WO_3 when heated at 550°C. The microspheres annealed at 350°C showed fast and the highest response to NO_2 gas (525 to 20 ppm) at 350°C (Liu et al. 2009). In 2010, Mukherjee investigated the gas sensing characteristics of cubic spinel-based nanocrystalline magnesium zinc ferrite powders. The n-type semiconducting magnesium–zinc ferrite exhibits reasonably good sensitivity towards various gases, including CO, H_2, CH_4, and N_2O. The sensors can be selective to hydrogen gas sensing by modulating the operating temperature. The sensitivity was measured at a constant gas concentration of ~ 1660 ppm. The maximum sensitivity was 19% for NO_2 and 43% for CO at 300°C (Mukherjee and Majumder 2010). Another work published in 2012 reported the improvement of a bare-Ga_2O_3 nanorod sensor for NO_2 by modifying Ga_2O_3-core/ZnO-shell nanorods fabricated by a two-step process comprising the thermal evaporation of GaN powders and atomic layer deposition of ZnO. The sensor showed a response of ~ 33% at a NO_2 concentration of 100 ppm at 300°C with a detection range of 10–1200 ppm (Jin et al. 2012). In 2017, Capone and coworkers reported a NO_2 and acetone gas sensing by using Pd-doped Fe_3O_4/γ-Fe_2O_3 nanoparticles, with an operating temperature of 200–500°C, with an enhanced response at 350°C for 5 ppm of NO_2 (Capone et al. 2017).

2.4 Nanomaterials in Outer-space Designs

Nowadays, daily achievements occur in the aerospace field at the governmental and private levels. Because of this, we can start thinking about a new era for the space race in the 21st century. A relevant aspect in this new industry directly links to optimization and reduction of production costs. Nanotechnology offers a variety of opportunities for efficient solutions, especially because of the link between size and available reaction surfaces.

The greater use of synthetic nanomaterials has increased its applicability in different fields. One of those fields is catalysis, which is key for the development of initial chemical products, medicines, and raw materials. One important area using catalysts under high temperatures is related to energetic materials (Zhang et al. 2019). The main example of the use of nanoparticles in high-temperature conditions is related to solid fuels, energetic materials whose decomposition speed gives useful energy to propel, orient, and, in general, control the direction of the spaceship or equipment in outer space. Solid propellant performs in such a way that once the decomposition reaction is initiated, it is not possible to stop it or control it. The advantage of solid fuels is their storage simplicity. They do not require other mobile mechanisms due to their great stability and useful life. Unlike liquid or gaseous propellants, solid propellants will not produce leaks or losses of material. In outer space, with no resistance, the pushing is more efficient with less used propellant (Blachowicz and Ehrmann 2020, Liu et al. 2019).

Catalysts play an important role in the reaction of propellants, either accelerating or sometimes decelerating the fuel decomposition. Composite solid propellants (CSP) are made by an oxidizing agent like ammonium perchlorate (AP), a metallic fuel like aluminum, and a binder (commonly polybutadiene acrylonitrile (PBAN), hydroxyl-terminated polybutadiene (HTPB) or polybutadiene acrylic acid (PDAA)) that helps to bind the oxidizing agent with the fuel. CSP has been widely used as the main energy source to propel modern rockets in military and space applications. The combustion speed of solid propellants has been considered a vital parameter for rocket engine design. The burning rate can be adapted using different components, oxidizing level of charge and particle size, and, more commonly, by adding adequate burning catalyst. The catalytic activity is a superficial phenomenon in which oxidation and reduction occur and by which reaction continues due to the large surface available. An important difference between CSP catalysts and traditional reactions is their full decomposition with the propellant and, therefore, the impossibility of reutilization.

Since the use of the first catalysts with application in CSP, there has been a continuous interest in the synthesis and application of new materials in special nanoparticles due to their large surface area and their high activity in most of the catalytic processes. Nowadays, the focus point of research in CSP is the nanostructured catalyst with improvements in physicochemical properties (Sharma et al. 2015a). A detailed comprehension of properties, burning, and nanocatalyzed propellants combustion is essential for the use in rocket engines (Budhwar et al. 2018). Many combustion catalysts have been researched in the past, and some are currently being used in modern formulations of propellants. Different proven nanoscale particles are compatible with ingredients of CSP, presenting an increase in combustion speed, which indicates that additives of metal oxides at the nanoscale can impact combustion behavior (Pang et al. 2016).

To this date, considerable research work about nanocatalysts has been conducted aiming to identify the best catalyst for potential advantages in combustion speed performance. Transition metal oxides nanoparticles are the most studied catalysts used in many applications on rocket engines. Ferric oxide, cobalt oxide, copper oxide, manganese oxide, chromium oxide, copper chromite, and ZnO nanoparticles have improved the decomposition process of solid composites propellants based on ammonium perchlorate (AP) (Babar and Malik 2015, Dave et al. 2016, Hortelano and Luis de la Fuente 2019). Other applications with nanoparticles such as CuZnO, CoZnO, and NiZnO have demonstrated an impact on the thermal decomposition from AP. CuZnO has shown greater catalytic action than the other nanocatalysts where the high-temperature decomposition of AP descended around 117°C. Results showed that AP activation energies with transition metal oxides could be promising candidates for CSP (Vara et al. 2019).

Ferric oxide (Fe_2O_3) has been one of the most common catalysts of AP, with particle sizes around 0.25–0.65 microns. It is widely used as a catalyst over other metal oxide materials such as ZnO, CuO, Al_2O_3, and SiO_2 due to its capacity to lower the high decomposing temperature of AP to lower temperatures oscillating between 320°C–380°C. It has been also demonstrated that nanoparticles of α-Fe_2O_3 have a catalytic effect over AP thermal decomposition, decreasing its low-temperature decomposition (LTD) at 14.4°C and its high-temperature decomposition (HTD) at 53.4°C (Cao et al. 2019). The same occurs when using iron nanoparticles as catalysts in CSP based on AP, increasing the combustion speed due to its decomposition at high temperatures (Styborski et al. 2015).

Synthesized nanoparticles of CuO have been researched through a green route using leaves extract of *Calotropis gigantean*, giving successful results for thermal catalytic decomposition of AP. These had a good catalytic effect over HTD of AP, changing the second peak of exothermic decomposition from 445°C to 330°C. These nanoparticles are efficient catalysts to improve the combustion speed of CSP, offering an advancement on the chemical method since its economical and environmentally friendly synthesis method that can be used for large-scale synthesis without using high pressure, energy, temperature, and toxic chemical products (Sharma et al. 2015b). Also, the catalysis mechanisms of CuO in complex with MoS_2 have been studied. Obtained results showed a decrease in the decomposition temperature of ammonium perchlorate from 417°C to 324°C, demonstrating once again that CuO nanoparticles are good candidates as catalysts for CSP (Hu et al. 2019).

Other research on nanoscale catalysts has prepared $CdFe_2O_4$ and Cd nanocrystals, showing a good catalytic effect over the thermal decomposition of AP, HTPB, and solid composite propellant, giving the latter an improvement in combustion speed. It was observed in the AP case that by adding Cd to AP, the first thermal peak changed from 310°C to 290°C, and the second thermal peak changed from 430°C to 400°C, with a final minor impact on the decomposition temperature of AP (Singh et al. 2010).

Another widely used burn rate catalyst is ferrocene (Fc), which has an extraordinary effect on enhancing CSP burn rates and energy release, along with good ignitability (Grythe and Hansen 2007, Subramanian 1999). However, simple Fc-based molecules and polymer derivatives have some significant disadvantages, such as easy migration of the catalyst to the surface of the solid propellant resulting in deterioration due to sublimation or evaporation and a high risk of unexpected explosion (Liu et al. 2015, Swarts et al. 1997). The linking of Fc to the hyper-branched polymer designs is an effective way to enhance the burn rate and reduce the migration tendency of ferrocene (Xiao et al. 2013). Xiao and coworkers evaluated Fc functionalized polyester and amine ester dendrimers as burn rate catalysts (Fengjuan et al. 2011, Xiao et al. 2012). Synthesized compounds showed a solid catalytic effect on the thermal decomposition of AP. On the other hand, Zain-ul-Abdin and coworkers tested Fc-terminated PAMAM dendrimers as burn rate catalysts of AP, getting a lower migration rate than Fc and catocene (Zain-ul-Abdin et al. 2017). Recently, this system was used to support the synthesis of copper nanoparticles, reducing the migration of ferrocene and having a catalytic effect on AP (Ríos et al. 2022), as Figure 3 shows.

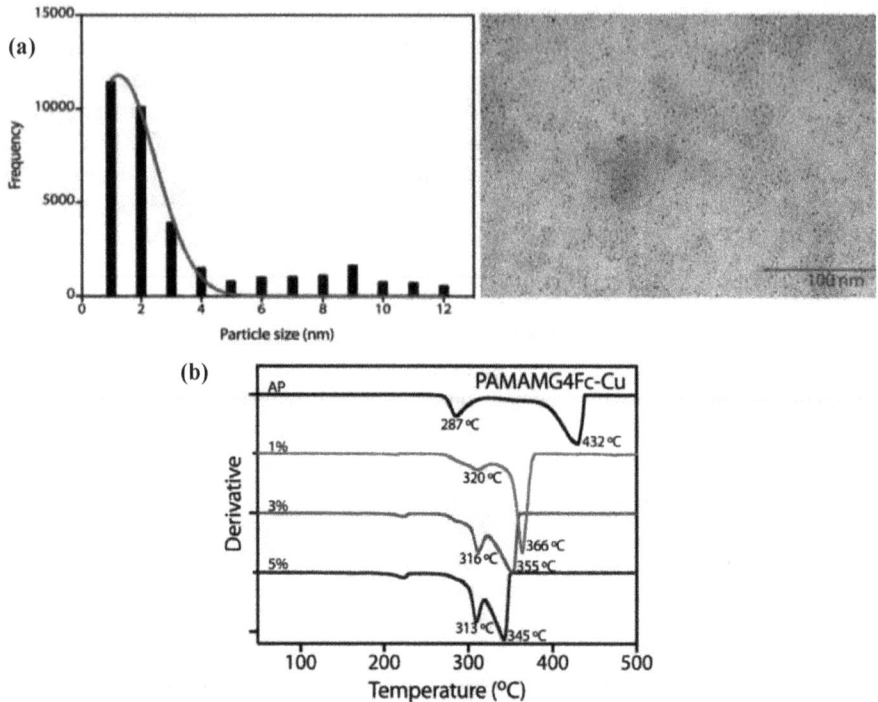

Figure 3. (a) TEM microphotograph of copper nanoparticles synthesized with PAMAM G4 dendrimers, and (b) the effect of copper nanoparticles stabilized on PAMAM G4-ferrocene on the decomposition of ammonium perchlorate. Figure adapted from (Ríos et al. 2022).

In search of new catalysts, Sabourin and collaborators analyzed graphene to improve performance in the combustion of CSP. This work highlighted the role of graphene in increasing the combustion speed of propellants, which catalyzes combustion reactions of fuel and energetically takes part, consuming itself without producing residual particles (Sabourin et al. 2009). The positive effects of graphene oxide in energetic materials are mainly based on higher thermal and electric conductivity and its large specific surface that impacts HTD and LTD of AP, speeding up the electronic transference and its decomposition in CSP (Melo et al. 2018). Other research has focused on the catalytic activity of copper nanoparticles supported on reduced graphene oxide. Graphene oxide and its derivatives have electronic rich groups that stabilize nanoparticles and avoid agglomeration. The results showed a decreased HTD of AP from 415°C to 318°C, accelerating the thermal decomposition of AP (Ríos et al. 2019). Quantum dots of reduced graphene oxide have also been tested, reducing the thermal decomposition of AP from 415°C to 392°C, showing a lower catalytic effect than graphene oxide (Melo et al. 2018). The catalytic activity of Co(OH)$_2$ nanoparticles supported by reduced graphene oxide positively affected the HTD of AP, decreasing its decomposition temperature to 292°C and increasing the energy release. This result is found among the best-reported values up to date, which suggests its potential application as catalysts in solid composites propellants (Abarca et al. 2020). It is worth mentioning that the development of almost all materials based on graphene derivatives has been informed during the last ten years. Many other compounds could be deposited on these materials to improve their performance as potential combustion catalysts or energetic additives for CSP.

An interesting application of nanotechnology in rocket engines based on liquid fuels was embedding a rocket engine combustion chamber with high thermal conductivity multiwall CNTs (MWCNTs) and diamond (D) particles using powder metallurgy techniques. The assays were unsuccessful due to the low thermal conductivity of the MWCNTs used in the tests. New studies are needed to clarify the potential of CNTs with higher thermal conductivity to produce hybrid materials to reduce mass, improve engine efficiency, and protect the engine from high pressure and temperature (Samareh and Siochi 2017).

Conclusion

The use of nanomaterials as additives in low quantities enhances and extends the traditional capabilities of materials; the reduction of wear translates into better mechanical performance. The modification of surfaces through the deposition of superhard nanocomposite thin films enhances mechanical and chemical resistance during the machining process. Furthermore, nanolubrication can extend the lifespan of moving parts according to the tribological characteristics of the surface. Researchers are still working on novel materials for machining operations, with superior capacities of strength and cost that can successfully replace today's best benchmark options. On the other hand, nanolubrication is still in debt to find a cost-effective nanoparticle/base fluid combination that is also eco-friendly, to reduce costs in a highly demanding industry, meeting hazard safety to the operators, and the increasing environmental regional regulations.

In order to reduce the risk through the monitoring of toxic substances, gas sensors based on nanoparticles have contributed to detecting different species at temperatures higher than 250°C. Increasing the operating range of the sensor, harmful substances can be monitored in more extreme conditions. However, there is still a necessity for designing new systems with better stability and detection limits.

Besides, nanoparticles as catalysts play an important role in the reaction of solid propellants, enhancing efficiency during the combustion process and decreasing the temperature of the reaction. Nanoscale catalysts are promising elements for current and future energetic applications.

Finally, great efforts have been made in the area of nanoparticles under extreme conditions, and must continue to obtain materials that lead to the highest possible productivity levels. As described here, it is a "thing of time" to explore new nanomaterials to satisfy the requirements under high-temperature conditions.

Acknowledgments

M.B.C. is grateful to FONDECYT Regular #1180023 for financial support.

References

Abarca, G., Ríos, P.L., Povea, P., Cerda-Cavieres, C., Morales-Verdejo, C., Arroyo, J.L. and Camarada, M.B. (2020). Nanohybrids of reduced graphene oxide and cobalt hydroxide (Co(OH)2|rGO) for the thermal decomposition of ammonium perchlorate. *RSC Advances*, 10(39): 23165–23172. doi:10.1039/D0RA02853C.

Afzal, A. (2019). β-Ga2O3 nanowires and thin films for metal oxide semiconductor gas sensors: Sensing mechanisms and performance enhancement strategies. *Journal of Materiomics*, 5(4): 542–557. doi:https://doi.org/10.1016/j.jmat.2019.08.003.

AL-Rjoub, A., Cavaleiro, A. and Fernandes, F. (2020). Influence of Ag alloying on the morphology, structure, mechanical properties, thermal stability and oxidation resistance of multilayered TiSiN/Ti (Ag) N films. *Materials and Design*, 108703.

Ali, M.K.A. and Xianjun, H. (2020a). Colloidal stability mechanism of copper nanomaterials modified by bis(2-ethylhexyl) phosphate dispersed in polyalphaolefin oil as green nanolubricants. *Journal of Colloid and Interface Science*, 578: 24–36. doi:https://doi.org/10.1016/j.jcis.2020.05.092.

Ali, M.K.A. and Xianjun, H. (2020b). Improving the heat transfer capability and thermal stability of vehicle engine oils using Al2O3/TiO2 nanomaterials. *Powder Technology*, 363: 48–58. doi:https://doi.org/10.1016/j.powtec.2019.12.051.

Amrita, M., Srikant, R.R. and Sitaramaraju, A.V. (2014). Performance Evaluation of nanographite-based cutting fluid in machining process. *Materials and Manufacturing Processes*, 29(5): 600–605. doi:10.1080/10426914.2014.893060.

Arunachalam, R., Mannan, M. and Spowage, A. (2004). Residual stress and surface roughness when facing age hardened Inconel 718 with CBN and ceramic cutting tools. *International Journal of Machine Tools and Manufacture*, 44(9): 879–887.

Aslantas, K., Ucun, I. and Cicek, A. (2012). Tool life and wear mechanism of coated and uncoated Al2O3/TiCN mixed ceramic tools in turning hardened alloy steel. *Wear*, 274: 442–451.

Babar, Z.-u.-d. and Malik, A.Q. (2015). An investigation of thermal decomposition kinetics of nano zinc oxide catalyzed composite propellant. *Combustion Science and Technology*, 187(8): 1295–1315. doi:10.1080/00102202.2015.1035375.

Bagheri, M., Khodadadi, A.A., Mahjoub, A.R. and Mortazavi, Y. (2015). Strong effects of gallia on structure and selective responses of Ga$_2$O$_3$–In$_2$O$_3$ nanocomposite sensors to either ethanol, CO or CH4. *Sensors and Actuators B: Chemical*, 220: 590–599. doi:https://doi.org/10.1016/j.snb.2015.06.007.

Behera, B. and Chandra, S. (2016). An innovative gas sensor incorporating ZnO–CuO nanoflakes in planar MEMS technology. *Sensors and Actuators B: Chemical*, 229: 414–424. doi:https://doi.org/10.1016/j.snb.2016.01.079.

Bhattacharyya, P., Basu, P.K., Mondal, B. and Saha, H. (2008). A low power MEMS gas sensor based on nanocrystalline ZnO thin films for sensing methane. *Microelectronics Reliability*, 48(11): 1772–1779. doi:https://doi.org/10.1016/j.microrel.2008.07.063.

Bhaumik, S., Maggirwar, R., Datta, S. and Pathak, S.D. (2018). Analyses of anti-wear and extreme pressure properties of castor oil with zinc oxide nano friction modifiers. *Applied Surface Science*, 449: 277–286. doi:https://doi.org/10.1016/j.apsusc.2017.12.131.

Blachowicz, T. and Ehrmann, A. (2020). 3D printed MEMS technology—Recent developments and applications. *Micromachines*, 11(4): 434. Retrieved from https://www.mdpi.com/2072-666X/11/4/434.

Bobzin, K. (2017). High-performance coatings for cutting tools. *CIRP Journal of Manufacturing Science and Technology*, 18: 1–9.

Bobzin, K., Brögelmann, T., Grundmeier, G., de Los Arcos, T., Wiesing, M. and Kruppe, N. (2017). A contribution to explain the mechanisms of adhesive wear in plastics processing by example of polycarbonate. *Surface and Coatings Technology*, 332: 464–473.

Bobzin, K., Brögelmann, T., Kruppe, N. and Carlet, M. (2019). Nanocomposite (Ti, Al, Cr, Si) N HPPMS coatings for high performance cutting tools. *Surface and Coatings Technology*, 378: 124857.

Budhwar, A.S., Gautam, A., More, P.V., Pant, C.S., Banerjee, S. and Khanna, P.K. (2018). Modified iron oxide nanoparticles as burn rate enhancer in composite solid propellants. *Vacuum*, 156: 483–491. doi:https://doi.org/10.1016/j.vacuum.2018.08.013.

Çalışkan, H., Erdoğan, A., Panjan, P., Gök, M. and Karaoğlanlı, A. (2013). Micro-abrasion wear testing of multilayer nanocomposite TiAlSiN/TiSiN/TiAlN hard coatings deposited on AISI H11 steel. *Materiali in Tehnologije/Materials and Technology*, 47(5): 563–568.

Cao, S.-b., Han, X.-g., Zhang, L.-l., Wang, J.-x., Luo, Y., Zou, H.-k. and Chen, J.-f. (2019). Facile and scalable preparation of α-Fe2O3 nanoparticle by high-gravity reactive precipitation method for catalysis of solid propellants combustion. *Powder Technology*, 353: 444–449. doi:https://doi.org/10.1016/j.powtec.2019.05.062.

Capone, S., Benkovicova, M., Forleo, A., Jergel, M., Manera, M.G., Siffalovic, P., ... Rella, R. (2017). Palladium/γ-Fe2O3 nanoparticle mixtures for acetone and NO2 gas sensors. *Sensors and Actuators B: Chemical*, 243: 895–903. doi:https://doi.org/10.1016/j.snb.2016.12.027.

Chen, H.-W., Chan, Y.-C., Lee, J.-W. and Duh, J.-G. (2011). Oxidation resistance of nanocomposite CrAlSiN under long-time heat treatment. *Surface and Coatings Technology*, 206(7): 1571–1576.

Chen, L., Paulitsch, J., Du, Y. and Mayrhofer, P.H. (2012). Thermal stability and oxidation resistance of Ti–Al–N coatings. *Surface and Coatings Technology*, 206(11-12): 2954–2960.

Chen, W., Yan, A., Meng, X., Wu, D., Yao, D. and Zhang, D. (2018). Microstructural change and phase transformation in each individual layer of a nano-multilayered AlCrTiSiN high-entropy alloy nitride coating upon annealing. *Applied Surface Science*, 462: 1017–1028. doi:https://doi.org/10.1016/j.apsusc.2018.07.106.

Chim, Y., Ding, X., Zeng, X. and Zhang, S. (2009). Oxidation resistance of TiN, CrN, TiAlN and CrAlN coatings deposited by lateral rotating cathode arc. *Thin Solid Films*, 517(17): 4845–4849.

Dave, P.N., Ram, P.N. and Chaturvedi, S. (2016). Transition metal oxide nanoparticles: Potential nano-modifier for rocket propellants. *Particulate Science and Technology*, 34(6): 676–680. doi:10.1080/02726351.2015.1112326.

Du, H., Datta, P., Griffin, D., Aljarany, A. and Burnell-Gray, J. (2003). Oxidation and sulfidation behavior of AlTiN-coated Ti–46.7 Al–1.9 W–0.5 Si intermetallic with CrN and NbN diffusion barriers at 850 C. *Oxidation of Metals*, 60(1-2): 29–46.

Fengjuan, X., Minmei, S., Lei, P., Yunjun, L. and Junchai, Z. (2011). Ferrocene end-cap hyperbranched poly(amine-ester): structure and catalytic performance for thermal decomposition of ammonium perchlorate. *Journal of Inorganic and Organometallic Polymers and Materials*, 21(1): 175–181.

Fuentes, G.G., Almandoz, E., Pierrugues, R., Martínez, R., Rodríguez, R.J., Caro, J. and Vilaseca, M. (2010). High temperature tribological characterisation of TiAlSiN coatings produced by cathodic arc evaporation. *Surface and Coatings Technology*, 205(5): 1368–1373.

Gekonde, H.O. and Subramanian, S. (2002). Tribology of tool–chip interface and tool wear mechanisms. *Surface and Coatings Technology*, 149(2-3): 151–160.

Ghosh, A., Maity, A., Banerjee, R. and Majumder, S.B. (2017). Volatile organic compound sensing using copper oxide thin films: Addressing the cross sensitivity issue. *Journal of Alloys and Compounds*, 692: 108–118. doi:10.1016/j.jallcom.2016.09.001.

Grythe, K.F. and Hansen, F.K. (2007). Diffusion rates and the role of diffusion in solid propellant rocket motor adhesion. *Journal of Applied Polymer Science*, 103(3): 1529–1538.

He, N., Li, H., Ji, L., Liu, X., Zhou, H. and Chen, J. (2016). High temperature tribological properties of TiAlSiN coatings produced by hybrid PVD technology. *Tribology International*, 98: 133–143.

Holubář, P., Jilek, M. and Šima, M. (1999). Nanocomposite nc-TiAlSiN and nc-TiN–BN coatings: Their applications on substrates made of cemented carbide and results of cutting tests. *Surface and Coatings Technology*, 120: 184–188.

Hörling, A., Hultman, L., Odén, M., Sjölén, J. and Karlsson, L. (2005). Mechanical properties and machining performance of Ti1–xAlxN-coated cutting tools. *Surface and Coatings Technology*, 191(2-3): 384–392.

Hortelano, C. and Luis de la Fuente, J. (2019). Chapter 9—New Developments in Composite Propellants Catalysis: From Nanoparticles to Metallo-Polyurethanes. pp. 363–388. *In*: Yan, Q.-L., He, G.-Q., Liu, P.-J. and Gozin, M. (eds.). *Nanomaterials in Rocket Propulsion Systems*. Elsevier.

Hu, Y., Yang, Y., Fan, R., Lin, K., Hao, D., Xia, D. and Wang, P. (2019). Enhanced thermal decomposition properties and catalytic mechanism of ammonium perchlorate over CuO/MoS2 composite. *Applied Organometallic Chemistry*, 33(9): e5060. doi:https://doi.org/10.1002/aoc.5060.

Hua, X., Sun, J., Zhang, P., Liu, K., Wang, R., Ji, J. and Fu, Y. (2016). Tribological properties of laser microtextured surface bonded with composite solid lubricant at high temperature. *Journal of Tribology*, 138(3). doi:10.1115/1.4032522.

Jeong, Y.J., Balamurugan, C. and Lee, D.W. (2016). Enhanced CO2 gas-sensing performance of ZnO nanopowder by La loaded during simple hydrothermal method. *Sensors and Actuators B: Chemical*, 229: 288–296. doi:https://doi.org/10.1016/j.snb.2015.11.093.

Jin, C., Park, S., Kim, H. and Lee, C. (2012). Ultrasensitive multiple networked Ga2O3-core/ZnO-shell nanorod gas sensors. *Sensors and Actuators B: Chemical*, 161(1): 223–228. doi:https://doi.org/10.1016/j.snb.2011.10.023.

Kalita, P., Malshe, A.P. and Rajurkar, K.P. (2012). Study of tribo-chemical lubricant film formation during application of nanolubricants in minimum quantity lubrication (MQL) grinding. *CIRP Annals*, 61(1): 327–330. doi:https://doi.org/10.1016/j.cirp.2012.03.031.

Kim, B.-J. and Kim, J.-S. (2013). Gas sensing characteristics of MEMS gas sensor arrays in binary mixed-gas system. *Materials Chemistry and Physics*, 138(1): 366–374. doi:https://doi.org/10.1016/j.matchemphys.2012.12.002.

Lai, C.-H., Tsai, M.-H., Lin, S.-J. and Yeh, J.-W. (2007). Influence of substrate temperature on structure and mechanical, properties of multi-element (AlCrTaTiZr)N coatings. *Surface and Coatings Technology*, 201(16): 6993–6998. doi:https://doi.org/10.1016/j.surfcoat.2007.01.001.

Lee, K., Hwang, Y., Cheong, S., Choi, Y., Kwon, L., Lee, J. and Kim, S.H. (2009). Understanding the role of nanoparticles in nano-oil lubrication. *Tribology Letters*, 35(2): 127–131. doi:10.1007/s11249-009-9441-7.

Li, J., Su, M., Li, G. and Liu, L. (2020). Atomic structure revolution and excellent performance improvement of composites induced by laser ultrafine-nano technology. *Composites Part B: Engineering*, 185: 107792. doi:https://doi.org/10.1016/j.compositesb.2020.107792.

Li, P., Chen, L., Wang, S.Q., Yang, B., Du, Y., Li, J. and Wu, M.J. (2013). Microstructure, mechanical and thermal properties of TiAlN/CrAlN multilayer coatings. *International Journal of Refractory Metals and Hard Materials*, 40: 51–57.

Lin, H.-J., Baltrus, J.P., Gao, H., Ding, Y., Nam, C.-Y., Ohodnicki, P. and Gao, P.-X. (2016). Perovskite nanoparticle-sensitized Ga2O3 nanorod arrays for CO detection at high temperature. *ACS Applied Materials and Interfaces*, 8(14): 8880–8887. doi:10.1021/acsami.6b01709.

Liu, B., Li, X., Yang, J. and Gao, G. (2019). Recent advances in MEMS-based microthrusters. *Micromachines*, 10(12): 818. Retrieved from https://www.mdpi.com/2072-666X/10/12/818.

Liu, X., Zhang, W., Zhang, G. and Gao, Z. (2015). Low-migratory ionic ferrocene-based burning rate catalysts with high combustion catalytic efficiency. *New Journal of Chemistry*, 39(1): 155–162.

Liu, Z., Miyauchi, M., Yamazaki, T. and Shen, Y. (2009). Facile synthesis and NO2 gas sensing of tungsten oxide nanorods assembled microspheres. *Sensors and Actuators B: Chemical*, 140(2): 514–519. doi:https://doi.org/10.1016/j.snb.2009.04.059.

Maheswaran, R. and Sunil, J. (2016). Effect of nano sized garnet particles dispersion on the viscous behavior of extreme pressure lubricant oil. *Journal of Molecular Liquids*, 223: 643–651. doi:https://doi.org/10.1016/j.molliq.2016.08.106.

Maity, A., Ghosh, A. and Majumder, S.B. (2015). Understanding the anomalous conduction behavior in 'n' type tungsten oxide thin film during hydrogen gas sensing: Kinetic analyses of conductance transients. *Sensors and Actuators B-Chemical*, 220: 949–957. doi:10.1016/j.snb.2015.06.038.

Mao, C., Huang, Y., Zhou, X., Gan, H., Zhang, J. and Zhou, Z. (2014). The tribological properties of nanofluid used in minimum quantity lubrication grinding. *The International Journal of Advanced Manufacturing Technology*, 71(5): 1221–1228. doi:10.1007/s00170-013-5576-7.

Meier, D.C., Semancik, S., Button, B., Strelcov, E. and Kolmakov, A. (2007). Coupling nanowire chemiresistors with MEMS microhotplate gas sensing platforms. *Applied Physics Letters*, 91(6): 063118. doi:10.1063/1.2768861.

Melo, J.P., Ríos, P.L., Povea, P., Morales-Verdejo, C. and Camarada, M.B. (2018). Graphene oxide quantum dots as the support for the synthesis of gold nanoparticles and their applications as new catalysts for the decomposition of composite solid propellants. *ACS Omega*, 3(7): 7278–7287. doi:10.1021/acsomega.8b00837.

Mitterer, C., Holler, F., Üstel, F. and Heim, D. (2000). Application of hard coatings in aluminium die casting—soldering, erosion and thermal fatigue behaviour. *Surface and Coatings Technology*, 125(1-3): 233–239.

Mukherjee, K. and Majumder, S.B. (2010). Reducing gas sensing behavior of nano-crystalline magnesium–zinc ferrite powders. *Talanta*, 81(4): 1826–1832. doi:https://doi.org/10.1016/j.talanta.2010.03.042.

Narayanan, R. and El-Sayed, M.A. (2005). Catalysis with transition metal nanoparticles in colloidal solution: Nanoparticle shape dependence and stability. *The Journal of Physical Chemistry B*, 109(26): 12663–12676. doi:10.1021/jp051066p.

Nguyen, T.D., Kim, S.K. and Lee, D.B. (2009). High-temperature oxidation of nano-multilayered TiAlCrSiN thin films in air. *Surface and Coatings Technology*, 204(5): 697–704.

PalDey, S. and Deevi, S. (2003). Single layer and multilayer wear resistant coatings of (Ti, Al) N: A review. *Materials Science and Engineering: A*, 342(1-2): 58–79.

Pang, W., De Luca, L.T., Fan, X., Maggi, F., Xu, H., Xie, W. and Shi, X. (2016). Effects of different nano-sized metal oxide catalysts on the properties of composite solid propellants. *Combustion Science and Technology*, 188(3): 315–328. doi:10.1080/00102202.2015.1083986.

Pogrebnjak, A., Yakushchenko, I., Bagdasaryan, A., Bondar, O., Krause-Rehberg, R., Abadias, G., ... Beresnev, V. (2014). Microstructure, physical and chemical properties of nanostructured (Ti–Hf–Zr–V–Nb) N coatings under different deposition conditions. *Materials Chemistry and Physics*, 147(3): 1079–1091.

Pourpasha, H., Zeinali Heris, S., Mahian, O. and Wongwises, S. (2020). The effect of multi-wall carbon nanotubes/turbine meter oil nanofluid concentration on the thermophysical properties of lubricants. *Powder Technology*, 367: 133–142. doi:https://doi.org/10.1016/j.powtec.2020.03.037.

Ramachandran, K., Navaneethakrishnan, P. and Sivaraja, M. (2020). The influence of nickel oxide nanoparticle dispersion on the thermo stability of lubricant oil. *International Journal of Nanoscience*, 19(01): 1850044.

Ríos, P.L., Araya-Durán, I., Bonardd, S., Arroyo, J.L., Povea, P. and Camarada, M.B. (2022). Ferrocene-modified dendrimers as support of copper nanoparticles: Evaluation of the catalytic activity for the decomposition of ammonium perchlorate. *Materials Today Chemistry*, 23: 100631. doi:https://doi.org/10.1016/j.mtchem.2021.100631.

Ríos, P.L., Povea, P., Cerda-Cavieres, C., Arroyo, J.L., Morales-Verdejo, C., Abarca, G. and Camarada, M.B. (2019). Novel *in situ* synthesis of copper nanoparticles supported on reduced graphene oxide and its application as a new catalyst for the decomposition of composite solid propellants. *RSC Advances*, 9(15): 8480–8489. doi:10.1039/C9RA00789J.

Rodriguez, R.S., O'Keefe, T.L., Froehlich, C., Lewis, R.E., Sheldon, T.R. and Haynes, C.L. (2021). Sensing food contaminants: Advances in analytical methods and techniques. *Analytical Chemistry*, 93(1): 23–40. doi:10.1021/acs.analchem.0c04357.

Rodríguez-Barrero, S., Fernández-Larrinoa, J., Azkona, I., López de Lacalle, L. and Polvorosa, R. (2016). Enhanced performance of nanostructured coatings for drilling by droplet elimination. *Materials and Manufacturing Processes*, 31(5): 593–602.

Ryu, H., Li, B., De Guise, S., McCutcheon, J. and Lei, Y. (2021). Recent progress in the detection of emerging contaminants PFASs. *Journal of Hazardous Materials*, 408: 124437. doi:https://doi.org/10.1016/j.jhazmat.2020.124437.

Sabourin, J.L., Dabbs, D.M., Yetter, R.A., Dryer, F.L. and Aksay, I.A. (2009). Functionalized graphene sheet colloids for enhanced fuel/propellant combustion. *ACS Nano*, 3(12): 3945–3954. doi:10.1021/nn901006w.

Samareh, J.A. and Siochi, E.J. (2017). Systems analysis of carbon nanotubes: Opportunities and challenges for space applications. *Nanotechnology*, 28(37): 372001. doi:10.1088/1361-6528/aa7c5a.

Sandstrom, D. and Hodowany, J. (1998). Modeling the physics of metal cutting in high-speed machining. *Machining Science and Technology*, 2(2): 343–353.

Shao, W., Ge, Z. and Song, Z. (2020). Bayesian just-in-time learning and its application to industrial soft sensing. *IEEE Transactions on Industrial Informatics*, 16(4): 2787–2798. doi:10.1109/TII.2019.2950272.

Sharma, J.K., Srivastava, P., Singh, G., Akhtar, M.S. and Ameen, S. (2015a). Catalytic thermal decomposition of ammonium perchlorate and combustion of composite solid propellants over green synthesized CuO nanoparticles. *Thermochimica Acta*, 614: 110–115. doi:https://doi.org/10.1016/j.tca.2015.06.023.

Sharma, N., Ojha, H., Bharadwaj, A., Pathak, D.P. and Sharma, R.K. (2015b). Preparation and catalytic applications of nanomaterials: A review. *RSC Advances*, 5(66): 53381–53403. doi:10.1039/C5RA06778B.

Sharmin, I., Gafur, M.A. and Dhar, N.R. (2020). Preparation and evaluation of a stable CNT-water based nano cutting fluid for machining hard-to-cut material. *SN Applied Sciences*, 2(4): 626. doi:10.1007/s42452-020-2416-x.

Shen, B., Shih, A.J. and Tung, S.C. (2008). Application of nanofluids in minimum quantity lubrication grinding. *Tribology Transactions*, 51(6): 730–737. doi:10.1080/10402000802071277.

Singh, G., Kapoor, I.P.S., Dubey, R. and Srivastava, P. (2010). Preparation, characterization and catalytic behavior of CdFe2O4 and Cd nanocrystals on AP, HTPB and composite solid propellants, Part: 79. *Thermochimica Acta*, 511(1): 112–118. doi:https://doi.org/10.1016/j.tca.2010.08.001.

Sirinakis, G., Siddique, R., Manning, I., Rogers, P.H. and Carpenter, M.A. (2006). Development and characterization of Au−YSZ surface plasmon resonance based sensing materials: High temperature detection of CO. *The Journal of Physical Chemistry B*, 110(27): 13508–13511. doi:10.1021/jp062760n.

Sofiah, A.G.N., Samykano, M., Shahabuddin, S., Kadirgama, K. and Pandey, A.K. (2020). An experimental study on characterization and properties of eco-friendly nanolubricant containing polyaniline (PANI) nanotubes blended in RBD palm olein oil. *Journal of Thermal Analysis and Calorimetry*. doi:10.1007/s10973-020-09891-6

Srikiran, S., Ramji, K., Satyanarayana, B. and Ramana, S.V. (2013). Investigation on turning of AISI 1040 steel with the application of nano-crystalline graphite powder as lubricant. *Proceedings of the Institution of Mechanical Engineers, Part C: Journal of Mechanical Engineering Science*, 228(9): 1570–1580. doi:10.1177/0954406213509612.

Styborski, J.A., Scorza, M.J., Smith, M.N. and Oehlschlaeger, M.A. (2015). Iron nanoparticle additives as burning rate enhancers in AP/HTPB composite propellants. *Propellants, Explosives, Pyrotechnics*, 40(2): 253–259. doi:https://doi.org/10.1002/prep.201400078.

Subramanian, K. (1999). Synthesis and characterization of poly (vinyl ferrocene) grafted hydroxyl-terminated poly (butadiene): A propellant binder with a built-in burn-rate catalyst. *Journal of Polymer Science Part A: Polymer Chemistry*, 37(22): 4090–4099.

Sui, X., Li, G., Qin, X., Yu, H., Zhou, X., Wang, K. and Wang, Q. (2016). Relationship of microstructure, mechanical properties and titanium cutting performance of TiAlN/TiAlSiN composite coated tool. *Ceramics International*, 42(6): 7524–7532.

Sun, X., Qiao, Y., Song, W., Ma, S. and Hu, C. (2013). High temperature tribological properties of modified nano-diamond additive in lubricating oil. *Physics Procedia*, 50: 343–347.

Sunil, J., Maheswaran, R., Vettumperumal, R. and Sadasivuni, K.K. (2019a). Experimental investigation on the thermal properties of NiO-nanofluids. *Journal of Nanofluids*, 8(7): 1577–1582.

Sunil, J., Vignesh, J., Vettumperumal, R., Maheswaran, R. and Raja, R.A.A. (2019b). The thermal properties of CaO-Nanofluids. *Vacuum*, 161: 383–388. doi:https://doi.org/10.1016/j.vacuum.2019.01.010.

Swarts, P.J., Immelman, M., Lamprecht, G.J., Greyling, S.E. and Swarts, J.C. (1997). Ferrocene derivatives as high burning rate catalysts in composite propellants. *South African Journal of Chemistry*, 50: 208–216.

Vamsi Krishna, P., Srikant, R.R. and Nageswara Rao, D. (2010). Experimental investigation on the performance of nanoboric acid suspensions in SAE-40 and coconut oil during turning of AISI 1040 steel. *International Journal of Machine Tools and Manufacture*, 50(10): 911–916. doi:https://doi.org/10.1016/j.ijmachtools.2010.06.001.

Vara, J.A., Dave, P.N. and Chaturvedi, S. (2019). The catalytic activity of transition metal oxide nanoparticles on thermal decomposition of ammonium perchlorate. *Defence Technology*, 15(4): 629–635. doi:https://doi.org/10.1016/j.dt.2019.04.002.

Vepřek, S. (1999). The search for novel, superhard materials. *Journal of Vacuum Science and Technology A: Vacuum, Surfaces, and Films*, 17(5): 2401–2420.

Veprek, S., Männling, H.-D., Jilek, M. and Holubar, P. (2004). Avoiding the high-temperature decomposition and softening of (Al1− xTix) N coatings by the formation of stable superhard NC-(Al1–xTix) N/a-Si3N4 nanocomposite. *Materials Science and Engineering: A*, 366(1): 202–205.

Wang, S.Q., Chen, L., Yang, B., Chang, K.K., Du, Y., Li, J. and Gang, T. (2010). Effect of Si addition on microstructure and mechanical properties of Ti–Al–N coating. *International Journal of Refractory Metals and Hard Materials*, 28(5): 593–596.

Wetchakun, K., Samerjai, T., Tamaekong, N., Liewhiran, C., Siriwong, C., Kruefu, V., ... Phanichphant, S. (2011). Semiconducting metal oxides as sensors for environmentally hazardous gases. *Sensors and Actuators B: Chemical*, 160(1): 580–591. doi:https://doi.org/10.1016/j.snb.2011.08.032.

Xiao, F., Feng, F., Li, L. and Zhang, D. (2013). Investigation on ultraviolet absorption properties, migration, and catalytic performances of ferrocene-modified hyper-branched polyesters. *Propellants, Explosives, Pyrotechnics*, 38(3): 358–365.

Xiao, F., Sun, X., Wu, X., Zhao, J. and Luo, Y. (2012). Synthesis and characterization of ferrocenyl-functionalized polyester dendrimers and catalytic performance for thermal decomposition of ammonium perchlorate. *Journal of Organometallic Chemistry*, 713: 96–103.

Xu, Y.X., Chen, L., Liu, Z.Q., Pei, F. and Du, Y. (2017). Improving thermal stability of TiSiN nanocomposite coatings by multilayered epitaxial growth. *Surface and Coatings Technology*, 321: 180–185.

Yu, D., Wang, C., Cheng, X. and Zhang, F. (2009). Microstructure and properties of TiAlSiN coatings prepared by hybrid PVD technology. *Thin Solid Films*, 517(17): 4950–4955.

Zain-ul-Abdin, L.W., Yu, H., Saleem, M., Akram, M., Khalid, H., Abbasi, N.M. and Yang, X. (2017). Synthesis of ethylene diamine-based ferrocene terminated dendrimers and their application as burning rate catalysts. *Journal of Colloid and Interface Science*, 487: 38–51.

Zarur, A.J. and Ying, J.Y. (2000). Reverse microemulsion synthesis of nanostructured complex oxides for catalytic combustion. *Nature*, 403(6765): 65–67. doi:10.1038/47450.

Zhang, D., Cao, C.-Y., Lu, S., Cheng, Y. and Zhang, H.-P. (2019). Experimental insight into catalytic mechanism of transition metal oxide nanoparticles on combustion of 5-Amino-1H-Tetrazole energetic propellant by multi kinetics methods and TG-FTIR-MS analysis. *Fuel*, 245: 78–88. doi:https://doi.org/10.1016/j.fuel.2019.02.007.

Zhou, C., Zhang, X., Tang, N., Fang, Y., Zhang, H. and Duan, X. (2020). Rapid response flexible humidity sensor for respiration monitoring using nano-confined strategy. *Nanotechnology*, 31(12): 125302. doi:10.1088/1361-6528/ab5cda.

Chapter 5

Nanomaterials under Corrosive Conditions

Andrés Ramírez* and Fabiola Pineda*

1. Introduction

The corrosion concept has a Latin origin from the *rodere* term, which means gnawing; thus, *corrodere* means gnawing in pieces. Nowadays, corrosion is defined as an irreversible interfacial reaction of any type of material with the environment, which provokes its gradual or total dissolution. This situation impedes that the material complies with the requirements for which it was designed and or selected; therefore, it raises serious concerns about conservation, safety, and economics. It is estimated that the cost of damages related to corrosion is 3 to 5 percent of industrialized countries' gross national product (GNP) (Elboujdaini 2007).

The variety of conditions and forms in which corrosion can be developed is immense, from domestic to extreme circumstances. Likewise, the type of damage produced by corrosion can be uniform or localized with the generation of corrosion products and voids or cracks in the material. The possible types of corrosion have been organized through mechanisms that take into consideration the typology of the attack, the environmental conditions, and mechanical stresses in the following groups (Groysman 2010):

- Uniform corrosion
- Localized corrosion (crevice, pitting, intergranular)
- Stress corrosion cracking
- Hot and high-temperature corrosion
- Microbiologically influenced corrosion

Regardless of the mechanism, the situation can be prevented by (i) modifications in the metallic materials to improve their corrosion resistance, (ii) modifications to

Centro de Nanotecnología Aplicada, Facultad de Ciencias, Ingeniería y Tecnología, Universidad Mayor, Camino la Pirámide 5750, 8580745-Santiago, Chile.
* Corresponding authors: andres.ramirez@umayor.cl; fabiola.pineda@umayor.cl

the environment to diminish their corrosivity, and (iii) modification of the interface between the metal and the environment with coatings to avoid their contact (Schütze 2002). These strategies can be approached from nano-science and -technology. Indeed, the use of nanomaterials specially designed to show better mechanical has been reported successfully; however, this has not always been accompanied by better corrosion and oxidation resistance. Hence, the use of nanoinhibitors and nanocoatings has received most of the attention due to their quality in enhancing the corrosion resistance and performance of several materials. Consequently, the following paragraphs describe the available information on coatings based on nanocomposites, their preparation methodology, and some protection mechanisms. In addition, this chapter describes the main nanostructures employed in corrosion protection in non-aqueous electrolytes and high temperatures.

2. Coating Methods and Corrosion Protection Mechanisms

Classic polymers are among the most widely used compounds for corrosion protection, mainly from the family of epoxides (EP) and Polybenzoxazine (PBZ). These materials have shown good mechanical and dielectric properties, and low water absorption, generating a hydrophobic surface that minimizes the possibility of corrosion, in addition to its low cost (Radhamani et al. 2020). Nevertheless, the use of polymers was further promoted in 1979, when MacDiarmid and Heeger developed a conductive organic material from polyacetylene, revolutionizing this area and giving rise to what we know today as conductive polymers (CPs) (Macdiarmid and Heeger 1979).

In this way, CPs have played a fundamental role as alternative materials to retard the corrosion process, the most used families being the polyanilines (Pani), polypyrroles (PPy), polythiophenes (PTh), and polyindoles (PIn), among others. In this way, in literature, we found films constituted by a single polymer (Aguirre et al. 2017, Yao et al. 2020) or two polymers obtained in the form of layer-by-layer (Tuken et al. 2004), or as co-polymers (Del Valle et al. 2020, Gopi et al. 2015), to generate a synergistic effect between the different corrosion protection properties. Some of the properties are its high environmental stability, low cost, the possibility of chemical and electrochemical synthesis, high conductivity, and redox properties, the latter being one of the most interesting properties; it should be remembered that this redox process, also known as doping/undoping, is based on the oxidation and partial reduction of the main chain. Therefore, when generating a potential change, it causes a variation in its oxidation state that, by electroneutrality law, allows the absorption and desorption of different ions (Zhou et al. 2020).

The protection mechanism generated by CPs on the surface of metals has not been fully elucidated; nonetheless, four types of hypotheses are described in the literature (Deshpande et al. 2014): (i) the deposited polymer induces the formation of metal oxide on the metal surface, (ii) the polymer is reduced by the release of dopant ion(s), protecting the metal surface, (iii) the polymer/metal interface generates an electric field that limits the electronic transfer of the metal to an oxidant, (iv) the polymer causes an increase in pH on the metal surface, preventing the entry of oxidative species and therefore the oxidation of the metal. In addition, for this last theory, the CP film has low porosity and high adhesion, which generates a good barrier effect and low mobility of O_2 in the metal/CPs interface, respectively.

Despite presenting these interesting properties, CPs have three important problems that affect their properties against the corrosion process; the first corresponds to a poor adhesion on the metal surface, mainly because this junction is principally governed by electrostatic forces (Figure 1A). Additionally, even though these films initially offer a barrier effect, their porosity generates an easy insertion of species that promote corrosion (Figure 1B). Finally, the ease of ion exchange between doping species and highly oxidizing species found in the environment, such as chloride ions, eventually generates localized corrosion that affects the metal surface (Figure 1C). For this reason, in the literature, it is not unusual to find studies using both types of polymers, providing a coating with better protection conditions.

The inclusion of nanostructured materials in CPs has been projected as an essential alternative to minimize the different named problems. However, to achieve a positive effect on corrosion protection, there must be a balance between the distribution and agglomeration of these nanostructures. That is to say, theoretically, the best conditions are presented through a high allocation and a low or no aggregation, which will limit the corrosive agent's mobility and diffusion to reach the surface of the protected metal (Radhamani et al. 2020, Ramezanzadeh and Attar 2011) (Figure 2A). On the other hand, the morphology of the nanostructures directly influences the new properties of the polymer by means such as its mechanical characteristics and gloss, among others (Deyab 2020), generated by the interactions between the different phases, which create a co-continuous network of the polymer chain and the various forms of the nanostructures such as 0D, 1D, 2D and 3D (Figure 2B) (Pourhashem et al. 2020).

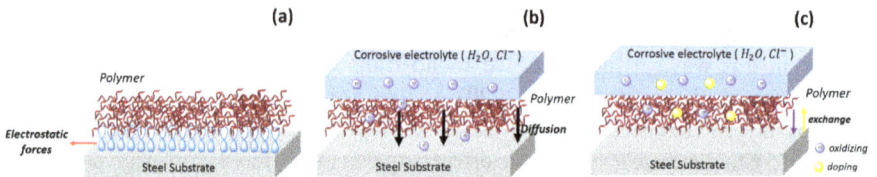

Figure 1. Schematic representation of the problems of depositing a polymer on steel (a) Electrostatic forces, (b) Porosity of the film, and (c) Ion exchange.

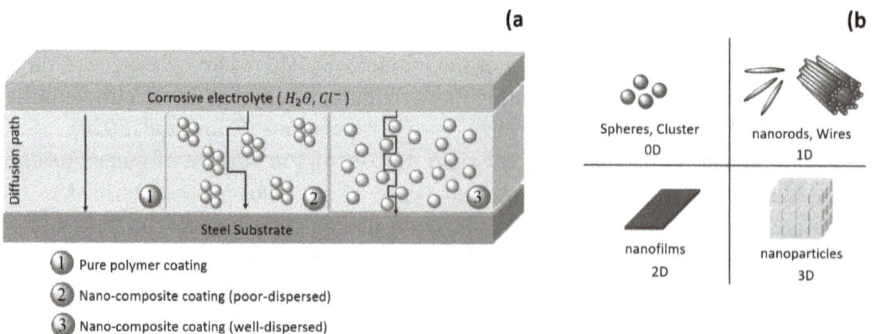

Figure 2. (a) Scheme of a protective film with and without nanostructures (Pourhashem et al. 2020) (b) Types of nanostructures.

In this context, it is observed that the contact angle increases in the presence of nanostructures, making the film more hydrophobic (Figure 3a). On the other hand, the presence of nanostructures over the metal surface is a very important factor because it improves the adhesion of the protective film, which prevents the generation of corrosion in the metal|polymer interface (Figure 3b). Additionally, Figure 3c shows the mechanism of the addition of nanostructures with a low oxidation potential, which, simulating the operation of a sacrificial anode, are rapidly oxidized when an oxidizing agent enters the polymeric matrix, preventing the metal surface from developing a corrosion product. Finally, nanostructured metal oxides, with high solubility in aqueous systems, release metal cations inside the polymeric matrix that diffuse the metal|polymer interface. Subsequently, after reaction with the aqueous medium and in the presence of a basic environment at the interface, they precipitate in the form of hydroxides, which eventually form a passive film on the surface of the metal, protecting it from oxidizing agents (Figure 3d).

The preparation stage of this type of film is considered a fundamental stage to guarantee the improvement of the properties against corrosion. It is possible to integrate these nanostructures during polymer shaping or after shaping. In this way, these three methods are generally described in the literature (Seidi et al. 2020, Yeasmin et al. 2021): (i) solution exfoliation (**SE**): the nanostructure and polymer are dispersed in specific solvents for each of the materials, which are subsequently mechanically mixed in order to obtain a polymer with nanostructure inside, and finally, the excess solvent is removed (Figure 4a), (ii) melt intercalation (**MI**), where a previously obtained polymer is used, and posteriorly its temperature is raised above its melting point, avoiding degradation, to later add the nanostructures to the already fused matrix, continuing with its constant stirring until the desired homogeneity

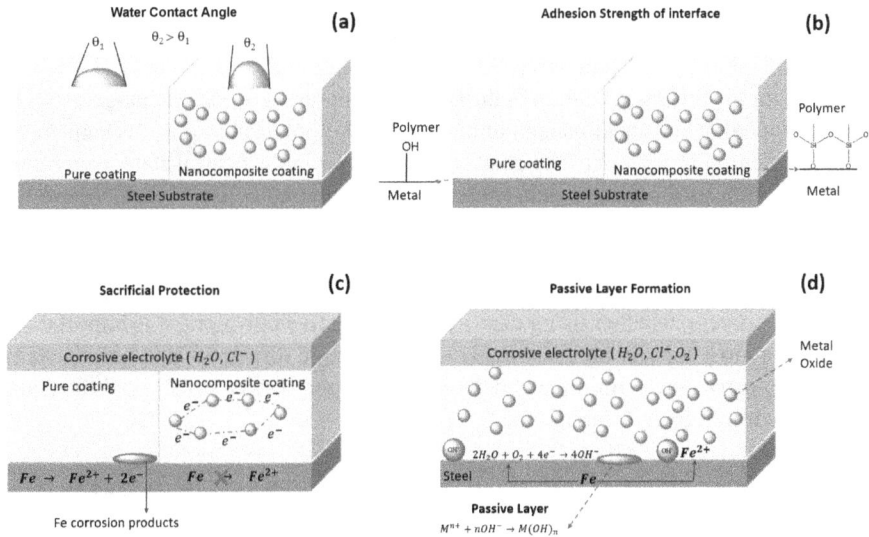

Figure 3. The corrosion protection mechanism of polymer coatings containing inorganic nanomaterials: (a) increasing water contact angle; (b) increasing adhesion strength at coating/metal interface; (c) sacrificial protection ; and (d) passive layer formation (Pourhashem et al. 2020).

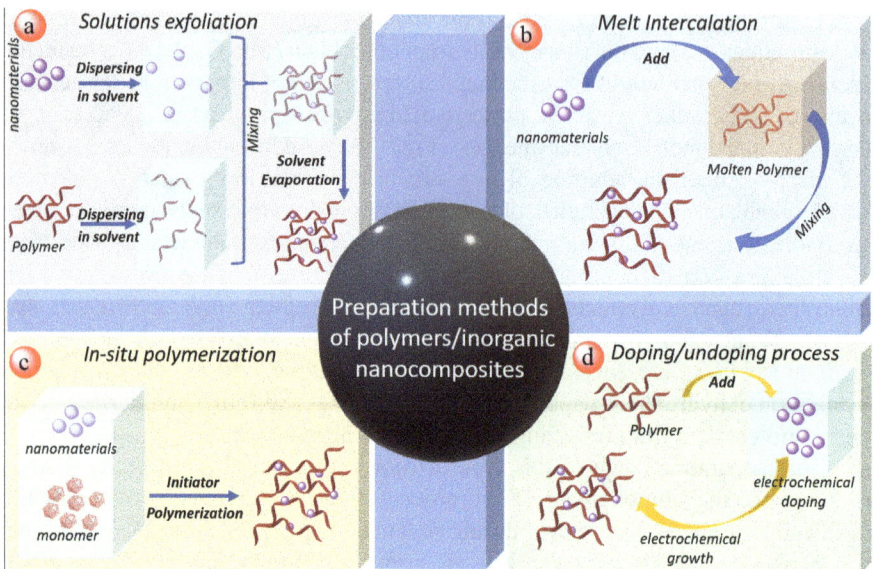

Figure 4. The methods for the preparation of polymer/inorganic nanocomposites: (a) solution exfoliation, (b) melt intercalation, (c) *in situ* polymerization, and (d) doping/undoping process. (Figure modified from ref (Pourhashem et al. 2020)).

between the polymer and the nanostructures is achieved (Pourhashem et al. 2020) (Figure 4b), and (iii) *in situ* polymerization (**ISP**), which uses the same solution where the precursor monomer is found, where the nanostructures are subsequently distributed randomly in the polymer matrix, the latter being obtained through various initiation techniques, depending on the monomer used (Figure 4c) (Wu et al. 2020). Consecutively, there is possibly a fourth method, which is rarely indicated and mainly based on conductive polymers' doping/undoping process (**DP**). The mechanism for this modification is through the generation of ion exchange between the doping anion and an anion containing the metal of interest. In this mechanism, an anodic potential is applied to the conductive polymer in the neutral state, generating the absorption of these ions; subsequently, a cathodic potential is used, where the anion containing the metal of interest is reduced to form a metallic nanostructure (Figure 4d) (Ramirez et al. 2017).

In summary, to produce a coating with improved properties that inhibit the corrosion process, whether using classical or conductive polymers, it is important to choose the most appropriate technique that favors good dispersion and homogeneity of the nanostructures, ultimately delivering a better barrier effect, stability, and greater hydrophobicity, among others.

3. Nanostructured Coatings and their Response in Natural Corrosion Environments

Different types of nanomaterials have been described in the literature, such as metal, metal oxides, carbon-based nanostructures (carbon nanotubes and graphene), and nanocapsules, which are discussed below, highlighting their primary functions as results.

Metallic nanostructures have been one of the first materials to be added to the different polymeric matrices, within which aluminum, titanium, and zinc stand out, among others. Initially, one of the first functions of these materials was to be used as sacrificial anodes due to their low oxidation potential compared to the metal to be protected, which are often different types of steel (Arianpouya et al. 2013). However, in recent studies, it has been observed that a barrier effect is generated after undergoing the oxidation of these metals (Kalendová et al. 2006), which preliminarily reduces the diffusion of corrosive agents such as chloride in the polymeric matrix (Zhang et al. 2007). On the other hand, it has been shown that the size of these nanostructures is important to generate a good barrier effect and allow homogeneity in the polymer matrix at the time of its application on the metal surface (Arman et al. 2013).

The amount of metal nanoparticles to be added to the polymeric matrix presents a critical concentration, where exceeding this concentration initiates agglomeration problems, as has been observed in the cases of Al and Zn. These agglomerations cause the matrix not to be able to bind strongly to metals, losing the uniformity of the coating, which finally facilitates the diffusion of corrosive agents towards the metal surface (Kalendová et al. 2006, Xue et al. 2007, Moazeni et al. 2013). Conversely, when the concentration of nanoparticles is ideal, a strong interaction between both species is achieved, translating into high resistance to corrosion (Zhang et al. 2006). Moreover, composites used for metal coating must present a good interaction between nanostructures and conductive polymers, fulfilling the purpose of adding superior anticorrosive properties and thus ensuring better protection against aggressive species, e.g., chlorides (Ramezanzadeh et al. 2017). The oxidation of metallic nanoparticles brings a second mechanism to protect against corrosion, where the formation of oxides and hydroxides occupy and fill the pores available in the polymeric matrix, which would act as a second barrier (Liu et al. 2012).

In this context, different types of metal oxides have also been studied, such as zeolites, SiO_2, TiO_2, ZnO, and Fe_2O_3, among others, which have been used mainly to fill the pores of the polymeric matrix, improving the protection performance through the reduction of corrosive agents (Eduok et al. 2017, Najjar et al. 2018, Shabani-Nooshabadi et al. 2018, Shi et al. 2017). In this way, it has been reported that the inclusion of nanostructured oxides increased the mechanical resistance of epoxy coatings (Ramezanzadeh and Attar 2011) and the thermal conductivity (Huang et al. 2020). At the same time, they reduce the water permeability which can decrease the mobility of oxidizing agents toward the metal surface (Mostafaei and Nasirpouri 2014), thus increasing the adhesion of the protective film on the metal surface (Jang et al. 2006, Pagotto et al. 2016), which has finally led to an improvement in corrosion protection and an increase in the useful life of the materials used, such as in marine vessels (Benea et al. 2018).

In contrast, a study in a saline environment of 304 steel coated with Polybenzoxazine and SiO_2 nanoparticles showed that the improvement in the adhesion of this on the metal surface; it can be attributed to three possible situations: (i) SiO_2 nanoparticles generate hydrophobicity that prevents the easy penetration of aqueous

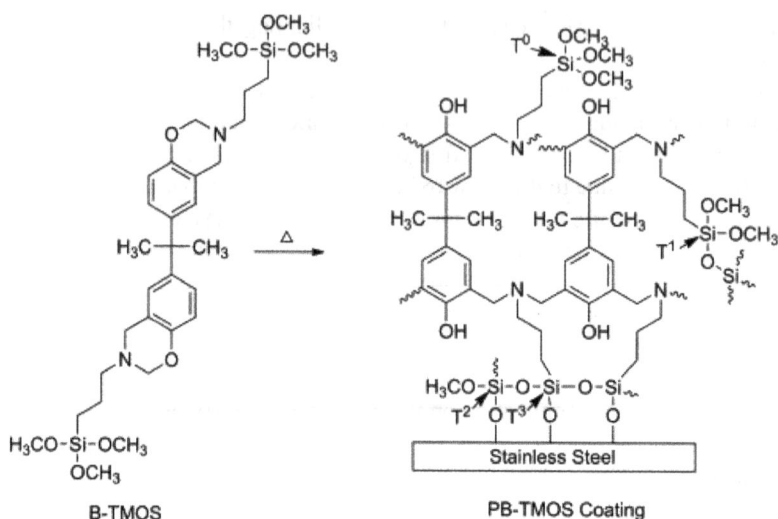

Figure 5. The formation process of PB-TMOS coating (Jang et al. 2006).

systems, (ii) water absorption by the coating is diminished due to the formation of a dual crosslinking network of polybenzoxazine and Si – O – Si or (iii) free alkoxy groups in the structure could reduce water transport to the substrate|coating interface, as shown in Figure 5 (Jang et al. 2006, Zhou et al. 2013).

Likewise, it has been described that the use of nanostructured TiO_2 notably improves corrosion protection in basic mediums through the increase in hydrophobicity, added to a self-protection effect that is activated when the surface suffers some scratches or blows. On the other hand, forming a layer on the metal surface generates an excellent barrier effect. Finally, the use of nanoparticles such as Fe_2O_3 shows significant improvements in protection against the corrosion process, both in acidic and basic environments, through a barrier effect by positioning itself between the grooves of the polymeric coating (Sumi et al. 2020).

The incorporation of carbon-based nanostructures, mainly oxidized graphene (GO) and carbon nanotubes (CNT), mixed with different polymeric matrices, has shown an enhancement in the corrosion protection properties. In this way, GO strengthens the mechanical properties of the coating through a labyrinth effect and high adhesion (Li et al. 2019, Radhamani et al. 2020). As long as CNTs have high electrical conductivity, broad mechanical resistance, and a significant lightness, that has allowed them to present interesting anticorrosive properties (Madhan Kumar and Gasem 2015a), generating an interesting synergistic effect with the polymer, which we will describe later.

In this way, like metallic nanoparticles, GO nanostructures have presented a critical concentration that ensures greater anti-corrosion property. In contrast, an excess of these nanostructures impairs the homogeneity of the film (Caldona et al. 2018). In the literature, a corrosion protection mechanism has been proposed, which consists of generating a barrier that, together with its flexibility, helps to waterproof gases, preventing the diffusion of oxygen and even that of water towards the metal surface, as shown in Figure 6 (Yu et al. 2014). However, one of the main problems

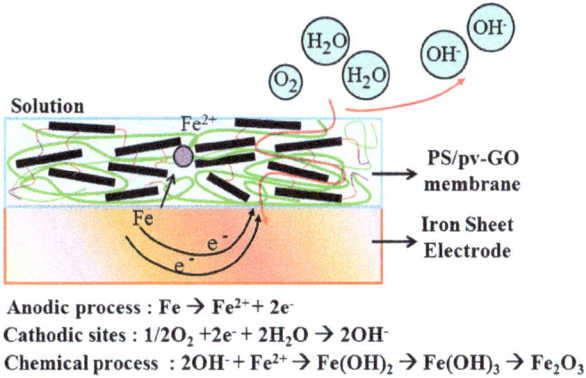

Anodic process : Fe → Fe^{2+} + 2e$^-$
Cathodic sites : $1/2O_2$ +2e$^-$ + $2H_2O$ → $2OH^-$
Chemical process : $2OH^-$ + Fe^{2+} → Fe(OH)$_2$ → Fe(OH)$_3$ → Fe$_2$O$_3$

Figure 6. Proposed mechanism of the improved properties for corrosion protection of the nanocomposites on the metal surface (Yu et al. 2014).

that GO has presented is that the large amounts of functional groups distributed on its surface, such as carboxyls and epoxides, among others, generate a bad dispersion of the nanostructures within the polymeric matrix, which generates permeabilization of water and therefore of corrosive ions towards the surface of the metal (Taheri et al. 2018, Zhou et al. 2019).

CNTs have presented a uniform dispersion in the polymeric matrix, which has generated an important hydrophobicity and mechanical resistance in humid environments, where the π-π interaction and the hydrogen bond maintains an electrical activity at a wide pH, being in a neutral and alkaline environment where it presents the best anticorrosive conditions (Madhan Kumar and Gasem 2015b, Qiu et al. 2017). Nevertheless, the inclusion of ionic liquids (IL) improved the corrosion protection between 1 and 3 orders of magnitude, where an increase is observed to be directly related to the immersion time. The protection mechanism is based on reducing the polymer that releases the ion from IL, which finally generates a layer of insoluble salts that, together with the CNTs, promotes a barrier effect on the metal surface (Souto and Soares 2020).

On the other hand, there are substances called corrosion inhibitors, which can be found mainly based on organic compounds, which tend to create a thin wall that prevents the corrosion process. Furthermore, when these structures are in nanometric dimensions, they induce an effective physisorption/chemisorption on the metal surface, generating a barrier effect (Zahidah et al. 2017). In addition, the inclusion of these nanostructures in the different polymeric matrices has provided the possibility of developing smart films, which have the property of self-healing against different surface damages, with a mechanism very similar to that shown in Figure 7a. This type of mechanism is activated in the face of changes in variables such as pressure, pH, temperature, and humidity, among others (Huang et al. 2018). Therefore, it will allow covering the metal surface again, preventing the corrosion of the material (Figure 7b–c).

Additionally, organic inhibitors can present aromatic rings within their chemical structure, such as nitrogen, oxygen, and sulfur, among others, which present only one pair of electrons available to promote corrosion protection. Thus, among the primary

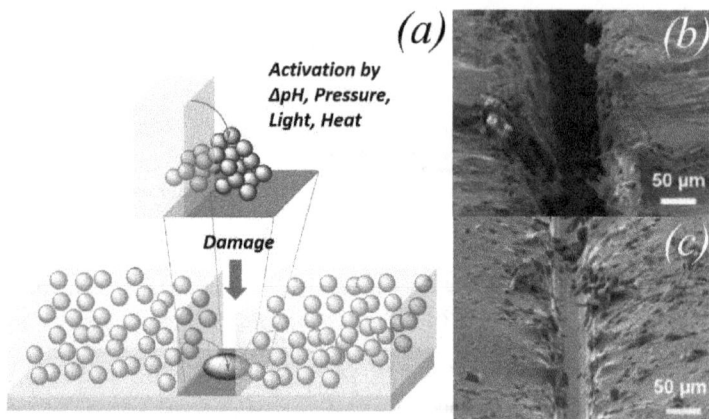

Figure 7. (a) Feed-back active containers release inhibitors in response to different stimuli. SEM images of the scribed region of (b) control coating and (c) self-healing coating after healing (Cho et al. 2009).

compounds used are epoxy and epoxy derivatives (Gu et al. 2016), isocyanurates (Nguyen et al. 2015), amines (Jin et al. 2012), and isocyanates (He et al. 2017). On the contrary, inorganic inhibitors have also focused attention on their intrinsic protective properties against corrosion, which can maintain homogeneity in the polymeric matrix, ensuring the barrier effect. Among the most used inorganic compounds are Zn, Al, and Mg (Zheludkevich et al. 2010), which are found inside nanocapsules, and are also made up of surfactants and polyelectrolytes.

Finally, it is important to point out that the participation of nanomaterials has had an important influence, enhancing their anticorrosive properties and, in addition, they managed to reduce the main problems that the most widely used polymeric coatings have presented. Therefore, it is important to continue investigating and deepening new nanomaterials, through which it is possible to reduce further problems that arise in new environments like extreme conditions.

4. Nanostructures in the Protection of Corrosion at High Temperatures

Lately, special attention has been focused on the corrosion phenomenon under extreme conditions of high pressure or temperature and direct contact with highly ionic electrolytes. In this sense emerges two concepts: hot and high-temperature corrosion. Hot corrosion (or hot oxidation in gaseous environments) is defined as "an accelerated corrosion of metal surfaces that results from the combined effect of oxidation and reactions with sulfur compounds and other contaminants, such as chlorides, to form a molten salt on a metal surface that fluxes, destroys, or disrupts the normal protective oxide" (International 2020). To date, there is no consensus on the definition of high-temperature corrosion (or high-temperature oxidation for gaseous environments). Still, there is agreement that it includes hot corrosion as it is a broader phenomenon identified in gaseous, molten salts, and liquid metals environments at temperatures above 400°C (Sandvik 2021).

The number of publications related to high-temperature corrosion has significantly increased over the past years, mainly because of the development and implementation of solar energy, particularly through concentrated solar power (CSP) plants. This is a relatively new approach to generating electricity from the sun's energy using mirrors that reflect sunlight in a reactor with molten salts inside, to be later stored in metal tanks allowing electricity generation from day and night. The 100% ionic nature of molten salts increases the metal components' corrosion risk drastically (Walczak et al. 2018). As was mentioned previously, corrosion can be prevented by modifying the metallic materials to improve their corrosion resistance, the environment to diminish their corrosivity, and the interface between the metal and the environment with coatings to avoid their contact (Schütze 2002). These strategies have already been studied from the nanoscience and nanotechnology point of view. Indeed, the use of nanomaterials specially designed to show better properties has been already reported successfully; however, this has not always been accompanied by better corrosion and oxidation resistance. Therefore, nanoinhibitors and nanocoatings have received most of the attention because they can enhance the corrosion resistance and performance of several materials.

In terms of materials, it is worth mentioning that the grains, grain boundaries, dislocations, stacking faults, deformation twins, voids, dispersoids, solute clusters, and precipitates correspond to the nanofeature of the microstructure that influences the corrosion resistance of materials (Ashby and Jones 2013). In this context, Raman and Gupta (Raman and Gupta 2009) studied the Fe–10%Cr alloy produced by ball milling, obtaining nanocrystalline (nc) and microcrystalline (mc) structures which were compared in terms of their oxidation performance being exposed to air at 300°C. The results of the gravimetric testing showed almost seven times higher mass gain for the mc alloy, with a parabolic rate law for nearly the entire exposure time; in contrast, the nc alloy only follows this rate law during the first 4 h. The difference in kinetic rate law indicates a different oxidation mechanism; in particular, the parabolic law is associated with oxide layer formation inversely proportional to the exposure time as a function of the diffusion of metal ions from the substrate toward the environment to react with components of it or the diffusion of oxygen ions through the oxide layer to interact with the metallic substrate (Khanna 2018). The authors attributed this to the fact that nc alloy presents higher grain boundary areas which corroded at the beginning of the exposure time. Later, this alloy develops a thinner multilayer structure with chromium oxide, presumably Cr_2O_3, as the inner layer and iron oxide as the outer layer. On the contrary, the mc alloy develops a mixed spinel internal layer type Fe-Cr oxide, which confers less oxidation resistance. This behavior has also been reported in alloys with aluminum developing a highly adherent Al_2O_3 that prevents oxidation (Pint et al. 2013, 2015, Rebak et al. 2018, Shi et al. 2020). The authors (Khanna 2018, Pint et al. 2013, 2015, Rebak et al. 2018, Shi et al. 2020) agree with nanoalloys' promising behavior since the initial fast diffusion that they presented required a significantly lower amount of passivating elements to form a completely protective oxide scale. Thus, it could be an approach to economize the manufacture of metal materials.

On the other hand, a corrosion inhibitor or nanoinhibitor is a chemical compound of micro or nanometric size added in a small amount to the environment that reduces or prevents the corrosion process (Sastri 2007). Nanoinhibitors are considered better because as the inhibitor's size decreases, the active surface area increases because there are more anchors sites to the metallic surface, allowing a superior inhibition efficiency. In this context, nanoparticles of different oxides have been studied as nanoinhibitors for hindering the corrosion process of metal materials exposed to molten salt in CSP plants. For this case, the inhibitor is added in lower percentages, usually 1 wt.%, to the environment, which is the molten salts. It is worth noting that the nanoparticles have been previously studied in molten salts since they increase their thermophysical properties. In fact, adding 1 wt.% of SiO_2 improves the thermal conductivity by 20%. However, no attention had been paid to their ability to inhibit corrosion until the study of Iyer (Iyer 2011). They were among the first groups studying this topic by adding SiO_2 to carbonate molten salt and monitoring the corrosion process of stainless steel by static gravimetric testing at 520°C. The authors postulated that the nanoparticles formed abducts – a chemical compound developed from the addition of two or more substances that correspond to molten salts and the NP for the case – that eventually turn into a passive layer reducing the corrosion rate. Indeed, this parameter was decreased by almost 50% in the presence of the nanoparticles. Unfortunately, the study did not include a complimentary analysis to confirm the nanoparticles' location and elucidate their inhibition mechanism. SiO_2 plus $Al2O_3$ nanoparticles were studied by Grosu et al. (Grosu et al. 2018), Fernández et al. (Fernández et al. 2019), and Nithiyanantham et al. (Nithiyanantham et al. 2019) as nanoinhibitors of the corrosion process of carbon, stainless, and carbon steel, respectively (in all cases by gravimetry at 310, 565 and 390°C). The results were not as promising as those of Iyer since the open-to-air atmosphere influenced the nanoparticles' coalescence and the generation of bubbles near them; thus, no significant enhancement of the corrosion resistance of the steels in the presence of the nanoparticles. Although these studies do not provide information about the mechanism related to nanoinhibitors, they clarify the conditions in which the benefits of their addition can be better. Nithiyanantham et al. (Nithiyanantham et al. 2019), in later research, studied the addition of 1 wt.% of TiO_2 nanoparticles to a nitrate molten salt in the corrosion process of carbon steel at 390°C. The results showed a significant diminishing in corrosion rate, which means that TiO_2 is an efficient nanoinhibitor. The authors confirm the theory of Iyer, which indicated that nanoparticles form a passive layer since they were found as the outer layer of the corrosion products forming an iron-titanium oxide. This layer hinders the diffusion of molten salt towards the steel surface and prevents the heterogeneous damage by the molten salts on the surface as well; in fact, no spallation or peel-off was observed.

Regarding nanocoatings, they are defined as either coating with a nanoscale thickness/grains/phases or coatings with a second phase in the nanosize range (Saji 2012). The first concept can be divided into nanostructured coatings, when the grains' size is lower than 100 nm and nanocomposited coatings, when two different materials are combined (Saji 2012). The second concept has been reported mainly by adding nanoparticles, which can improve the coating's adhesion, prevent blistering and delamination and enhance the barrier effect to avoid corrosion (Yeganeh et al. 2020). To date, the use of nanocomposited coatings for extreme conditions has not been reported since these harsh conditions demand thick films and highly dense structures. Nevertheless, one good approach is given by Encinas-Sánchez et al. (Encinas-Sánchez et al. 2018), who studied a nanocoating of ZrO_2 doped with 3 mol% of Y_2O_3 obtained by the sol-gel method on low-alloy steel exposed to nitrate molten salt at 500°C. The results were promising since the behavior of the coated steel is comparable with stainless steel, which is significantly expensive. Moreover, the authors analyzed the environmental impact through three variables: the cumulative energy demand, climate change, and water resource depletion for the coated low-alloy steel and stainless steel. In all the categories, the coated low-alloy steel was better than stainless steel, which indicates that this type of nanocoating contributes to reducing corrosion and reducing environmental impact, which is why it is highly recommended for massive use.

Conclusions

Corrosion is an inevitable process that causes severe damage to different structures. Thus, it is necessary to do more research and investment in protection to increase materials' useful life. In the case of coatings, it is necessary to ensure good adherence to the material for a good protection, which can be obtained by adding nanomaterials over the surface before the deposition of the coating. Also, the coating can be enhanced by adding nanomaterials inside, obtaining a smart coating that repairs itself. However, the methodology for adding the nanomaterials is still challenging. On the other hand, corrosion inhibitors are mainly recommended for several systems, i.e., small, and closed systems. Finally, a perspective on the relationship between nanomaterials and corrosion protection is essential to elucidate the mechanism and design strategies to improve corrosion resistance.

Acknowledgments

The authors acknowledge the financial support of ANID-Chile through Fondecyt, grant number Project N° 11190995 and N° 11200388.

References

Aguirre, J., Daille, L., Fischer, D.A., Galarce, C., Pizarro, G., Vargas, I. and Armijo, F. (2017). Study of poly(3,4-ethylendioxythiphene) as a coating for mitigation of biocorrosion of AISI 304 stainless steel in natural seawater. *Progress in Organic Coatings*, 113: 175–184. doi: 10.1016/j.porgcoat.2017.09.009.

Arianpouya, N., Shishesaz, M., Arianpouya, M. and Nematollahi, M. (2013). Evaluation of synergistic effect of nanozinc/nanoclay additives on the corrosion performance of zinc-rich polyurethane nanocomposite coatings using electrochemical properties and salt spray testing. *Surface and Coatings Technology*, 216: 199–206. doi: https://doi.org/10.1016/j.surfcoat.2012.11.036.

Arman, S.Y., Ramezanzadeh, B., Farghadani, S., Mehdipour, M. and Rajabi, A. (2013). Application of the electrochemical noise to investigate the corrosion resistance of an epoxy zinc-rich coating loaded with lamellar aluminum and micaceous iron oxide particles. *Corrosion Science*, 77: 118–127. doi: https://doi.org/10.1016/j.corsci.2013.07.034.

Ashby, M.F. and Jones, D.R.H. (2013). Chapter 1—Metals. pp. 3–14. *In*: Ashby, M.F. and Jones, D.R.H. (eds.). *Engineering Materials 2 (Fourth Edition)*. Boston: Butterworth-Heinemann.

Benea, L., Mardare, L. and Simionescu, N. (2018). Anticorrosion performances of modified polymeric coatings on E32 naval steel in sea water. *Progress in Organic Coatings*, 123: 120–127. doi: https://doi.org/10.1016/j.porgcoat.2018.06.020.

Caldona, E.B., de Leon, A.C.C., Mangadlao, J.D., Lim, K.J.A., Pajarito, B.B. and Advincula, R.C. (2018). On the enhanced corrosion resistance of elastomer-modified polybenzoxazine/graphene oxide nanocomposite coatings. *Reactive and Functional Polymers*, 123: 10–19. doi: https://doi.org/10.1016/j.reactfunctpolym.2017.12.004.

Cho, S.H., White, S.R. and Braun, P.V. (2009). Self-healing polymer coatings. *Advanced Materials*, 21(6): 645–649. doi: https://doi.org/10.1002/adma.200802008.

Del Valle, M.A., Mieres, E., Motheo, A. and Ramirez, A.M.R. (2020). Effect of the monomers ratio in the electrosynthesis of poly(Aniline-Co-O-Methoxyaniline) on steel corrosion protection. *Journal of the Chilean Chemical Society*, 65(4): 4998–5003.

Deshpande, P.P., Jadhav, N.G., Gelling, V.J. and Sazou, D. (2014). Conducting polymers for corrosion protection: A review. *Journal of Coatings Technology and Research*, 11(4): 473–494. doi: 10.1007/s11998-014-9586-7.

Deyab, M.A. (2020). Anticorrosion properties of nanocomposites coatings: A critical review. *Journal of Molecular Liquids*, 313. doi: Artn 11353310.1016/J.Molliq.2020.113533.

Eduok, U., Jossou, E. and Szpunar, J. (2017). Enhanced surface protective performance of chitosanic hydrogel via nano-CeO2 dispersion for API 5L X70 alloy: Experimental and theoretical investigations of the role of CeO2. *Journal of Molecular Liquids*, 241: 684–693. doi: https://doi.org/10.1016/j.molliq.2017.06.058.

Elboujdaini, V.S.S.E.G.M. (2007). Corrosion economics and corrosion management. *Corrosion Prevention and Protection*, pp. 311–328.

Encinas-Sánchez, V., Batuecas, E., Macías-García, A., Mayo, C., Díaz, R. and Pérez, F.J. (2018). Corrosion resistance of protective coatings against molten nitrate salts for thermal energy storage and their environmental impact in CSP technology. *Solar Energy*, 176: 688–697. doi: https://doi.org/10.1016/j.solener.2018.10.083.

Fernández, A.G., Muñoz-Sánchez, B., Nieto-Maestre, J. and García-Romero, A. (2019). High temperature corrosion behavior on molten nitrate salt-based nanofluids for CSP plants. *Renewable Energy*, 130: 902–909. doi: https://doi.org/10.1016/j.renene.2018.07.018.

Gopi, D., Karthikeyan, P., Kavitha, L. and Surendiran, M. (2015). Development of poly(3,4-ethylenedioxythiophene-co-indole-5-carboxylic acid) co-polymer coatings on passivated low-nickel stainless steel for enhanced corrosion resistance in the sulphuric acid medium. *Applied Surface Science*, 357: 122–130. doi: 10.1016/j.apsusc.2015.09.001.

Grosu, Y., Udayashankar, N., Bondarchuk, O., González-Fernández, L. and Faik, A. (2018). Unexpected effect of nanoparticles doping on the corrosivity of molten nitrate salt for thermal energy storage. *Solar Energy Materials and Solar Cells*, 178: 91–97. doi: https://doi.org/10.1016/j.solmat.2018.01.002.

Groysman, A. (2010). Corrosion for everybody. *Corrosion for Everybody*, 1–368. doi: 10.1007/978-90-481-3477-9.

Gu, J., Yang, X., Li, C. and Kou, K. (2016). Synthesis of cyanate ester microcapsules via solvent evaporation technique and its application in epoxy resins as a healing agent. *Industrial and Engineering Chemistry Research*, 55(41): 10941–10946. doi: 10.1021/acs.iecr.6b03093.

He, Z., Jiang, S., Li, Q., Wang, J., Zhao, Y. and Kang, M. (2017). Facile and cost-effective synthesis of isocyanate microcapsules via polyvinyl alcohol-mediated interfacial polymerization and their application in self-healing materials. *Composites Science and Technology*, 138: 15–23. doi: https://doi.org/10.1016/j.compscitech.2016.11.004.

Huang, W.F., Xiao, Y.L., Huang, Z.J., Tsui, G.C.P., Yeung, K.W., Tang, C.Y. and Liu, Q. (2020). Super-hydrophobic polyaniline-TiO2 hierarchical nanocomposite as anticorrosion coating. *Materials Letters*, 258: 126822. doi: https://doi.org/10.1016/j.matlet.2019.126822.

Huang, Z., Gurney, R.S., Wang, T. and Liu, D. (2018). Environmentally durable superhydrophobic surfaces with robust photocatalytic self-cleaning and self-healing properties prepared via versatile film deposition methods. *Journal of Colloid and Interface Science*, 527: 107–116. doi: https://doi.org/10.1016/j.jcis.2018.05.004.

International, A. (2020). ASTM NACEASTM G193-12d "Standard Terminology and Acronyms Relating to Corrosion. West Conshohocken, PA.

Iyer, A.P. (2011). *The Effect of Silica Nanoparticles on Corrosion of Steel by Molten Carbonate Eutectics.*

Jang, J., Ha, J. and Lim, B. (2006). Synthesis and characterization of monodisperse silica–polyaniline core–shell nanoparticles. *Chemical Communications*, (15): 1622–1624. doi: 10.1039/B600167J.

Jin, H., Mangun, C.L., Stradley, D.S., Moore, J.S., Sottos, N.R. and White, S.R. (2012). Self-healing thermoset using encapsulated epoxy-amine healing chemistry. *Polymer*, 53(2): 581–587. doi: https://doi.org/10.1016/j.polymer.2011.12.005.

Kalendová, A., Kalenda, P. and Veselý, D. (2006). Comparison of the efficiency of inorganic nonmetal pigments with zinc powder in anticorrosion paints. *Progress in Organic Coatings*, 57(1): 1–10. doi: https://doi.org/10.1016/j.porgcoat.2006.05.015.

Khanna, A.S. (2018). Chapter 6—High-temperature oxidation. pp. 117–132. *In*: Kutz, M. (ed.). *Handbook of Environmental Degradation of Materials (Third Edition)*. William Andrew Publishing.

Li, Y.Y., Xu, Y.T., Wang, S.C., Wang, H.C., Li, M. and Dai, L.Z. (2019). Preparation of graphene/polyaniline nanocomposite by *in situ* intercalation polymerization and their application in anti-corrosion coatings. *High Performance Polymers*, 31(9-10): 1226–1237. doi: 10.1177/0954008319839442.

Lili XUE, L.X. and Qingfen, L.I. (2007). Effect of nano Al pigment on the anti corrosive performance of waterborne epoxy coatings. *J. Mater. Sci. Technol.*, 23(04): 563–567.

Liu, J.H., Zhan, Z.W., Li, S.M. and Yu, M. (2012). Corrosion resistance of waterborne epoxy coating pigmented by nano-sized aluminium powder on steel. *Journal of Central South University*, 19(1): 46–54. doi: 10.1007/s11771-012-0971-z.

Macdiarmid, A.G. and Heeger, A.J. (1979). Organic metals and semiconductors—Polyacetylene, (Ch)X and its derivatives. *Journal of Electronic Materials*, 8(5): 718–718.

Madhan Kumar, A. and Gasem, Z.M. (2015a). Effect of functionalization of carbon nanotubes on mechanical and electrochemical behavior of polyaniline nanocomposite coatings. *Surface and Coatings Technology*, 276: 416–423. doi: https://doi.org/10.1016/j.surfcoat.2015.06.036.

Madhan Kumar, A. and Gasem, Z.M. (2015b). *In situ* electrochemical synthesis of polyaniline/f-MWCNT nanocomposite coatings on mild steel for corrosion protection in 3.5% NaCl solution. *Progress in Organic Coatings*, 78: 387–394. doi: https://doi.org/10.1016/j.porgcoat.2014.07.009.

Moazeni, N., Mohamad, Z., Faisal, N.L.I., Tehrani, M.A. and Dehbari, N. (2013). Anticorrosion epoxy coating enriched with hybrid nanozinc dust and halloysite nanotubes. *Journal of Applied Polymer Science*, 130(2): 955–960. doi: https://doi.org/10.1002/app.39239.

Mostafaei, A. and Nasirpouri, F. (2014). Epoxy/polyaniline–ZnO nanorods hybrid nanocomposite coatings: Synthesis, characterization and corrosion protection performance of conducting paints. *Progress in Organic Coatings*, 77(1): 146–159. doi: https://doi.org/10.1016/j.porgcoat.2013.08.015.

Najjar, R., Katourani, S.A. and Hosseini, M.G. (2018). Self-healing and corrosion protection performance of organic polysulfide@urea-formaldehyde resin core-shell nanoparticles in epoxy/PANI/ZnO nanocomposite coatings on anodized aluminum alloy. *Progress in Organic Coatings*, 124: 110–121. doi: https://doi.org/10.1016/j.porgcoat.2018.08.015.

Nguyen, L.-T.T., Hillewaere, X.K.D., Teixeira, R.F.A., van den Berg, O. and Du Prez, F.E. (2015). Efficient microencapsulation of a liquid isocyanate with *in situ* shell functionalization. *Polymer Chemistry*, 6(7): 1159–1170. doi: 10.1039/C4PY01448K.

Nithiyananthama, U., Grosu, Y., Anagnostopoulos, A., Carbó-Argibay, E., Bondarchuk, O., González-Fernández, L., Zaki, A., Igartua, J.M., Navarro, M.E., Ding, Y. and Faika, A. (2019). Nanoparticles as a high-temperature anticorrosion additive to molten nitrate salts for concentrated solar power. *Solar Energy Materials and Solar Cells*, 203: 110171. doi: https://doi.org/10.1016/j.solmat.2019.110171.

Nithiyantham, U., Grosu, Y., González-Fernández, L., Zaki, A., Igartua, J.M. and Faik, A. (2019). Corrosion aspects of molten nitrate salt-based nanofluids for thermal energy storage applications. *Solar Energy*, 189: 219–227. doi: https://doi.org/10.1016/j.solener.2019.07.050.

Pagotto, J.F., Recio, F.J., Motheo, A.J. and Herrasti, P. (2016). Multilayers of PAni/n-TiO2 and PAni on carbon steel and welded carbon steel for corrosion protection. *Surface and Coatings Technology*, 289: 23–28. doi: https://doi.org/10.1016/j.surfcoat.2016.01.046.

Pint, B.A., Terrani, K.A., Brady, M.P., Cheng, T. and Keiser, J.R. (2013). High temperature oxidation of fuel cladding candidate materials in steam–hydrogen environments. *Journal of Nuclear Materials*, 440(1): 420–427. doi: https://doi.org/10.1016/j.jnucmat.2013.05.047.

Pint, B.A., Unocic, K.A. and Terrani, K.A. (2015). Effect of steam on high temperature oxidation behaviour of alumina-forming alloys. *Materials at High Temperatures*, 32(1-2): 28–35. doi: 10.117 9/0960340914Z.00000000058.

Pourhashem, S., Saba, F., Duan, J.Z., Rashidi, A., Guan, F., Nezhad, E.G. and Hou, B.R. (2020). Polymer/ inorganic nanocomposite coatings with superior corrosion protection performance: A review. *Journal of Industrial and Engineering Chemistry*, 88: 29–57. doi: 10.1016/j.jiec.2020.04.029.

Qiu, G., Zhu, A. and Zhang, C. (2017). Hierarchically structured carbon nanotube–polyaniline nanobrushes for corrosion protection over a wide pH range. *Rsc Advances*, 7(56): 35330–35339. doi: 10.1039/ C7RA05235A.

Radhamani, A.V., Lau, H.C. and Ramakrishna, S. (2020). Nanocomposite coatings on steel for enhancing the corrosion resistance: A review. *Journal of Composite Materials*, 54(5): 681–701. doi: Artn 0021 99831985780710.1177/0021998319857807.

Raman, R.K.S. and Gupta, R.K. (2009). Oxidation resistance of nanocrystalline vis-à-vis microcrystalline Fe–Cr alloys. *Corrosion Science*, 51(2): 316–321. doi: https://doi.org/10.1016/j.corsci.2008.10.020.

Ramezanzadeh, B. and Attar, M.M. (2011). Studying the effects of micro and nano sized ZnO particles on the corrosion resistance and deterioration behavior of an epoxy-polyamide coating on hot-dip galvanized steel. *Progress in Organic Coatings*, 71(3): 314–328. doi: 10.1016/j.porgcoat.2011.03.026.

Ramezanzadeh, B., Moghadam, M.H.M., Shohani, N. and Mandavian, M. (2017). Effects of highly crystalline and conductive polyaniline/graphene oxide composites on the corrosion protection performance of a zinc-rich epoxy coating. *Chemical Engineering Journal*, 320: 363–375. doi: 10.1016/j.cej.2017.03.061.

Ramirez, M.R.A., del Valle, M.A., Armijo, F., Diaz, F.R., Pardo, M.A. and Ortega, E. (2017). Enhancement of electrodes modified by electrodeposited PEDOT-nanowires with dispersed Pt nanoparticles for formic acid electro-oxidation. *Journal of Applied Polymer Science*, 134(16). doi: Artn 4472310.1002/ App.44723.

Rebak, R.B., Gupta, V.K. and Larsen, M. (2018). Oxidation characteristics of two FeCrAl alloys in air and steam from 800°C to 1300°C. *Jom*, 70(8): 1484–1492. doi: 10.1007/s11837-018-2979-9.

Saji, V.S. (2012). 1—The impact of nanotechnology on reducing corrosion cost. pp. 3–15. *In*: Saji, V.S. and Cook, R. (eds.). *Corrosion Protection and Control Using Nanomaterials*. Woodhead Publishing.

Sandvik. (2021). High-Temperature Corrosion. Retrieved May 29, 2021, 2021, from https://www. materials.sandvik/en/materials-center/corrosion/high-temperature-corrosion/.

Sastri, V.S., Ghali, E. and Elboujdaini, M. (2007). Practical solutions. *Corrosion Prevention and Protection* (pp. 461–551).

Schütze, M. (2002). Corrosion Books: Handbook of Corrosion Engineering. By Pierre R. Roberge - Materials and Corrosion 4/2002. *Materials and Corrosion*, 53(4): 284–284. doi: https://doi. org/10.1002/1521-4176(200204)53:4<284::AID-MACO1111284>3.0.CO;2-8.

Seidi, F., Jouyandeh, M., Taghizadeh, M., Taghizadeh, A., Vahabi, H., Habibzadeh, S., Formela, K. and Saeb, M.R. (2020, Jun). Metal-organic framework (mof)/epoxy coatings: A review. *Materials (Basel)*, 26; 13(12): 2881. Doi: 10.3390/ma13122881.

Shabani-Nooshabadi, M., Allahyary, E. and Jafari, Y. (2018). Enhanced anti-corrosive properties of electrosynthesized polyaniline/zeolite nanocomposite coatings on steel. *Journal of Nanostructures*, 8(2): 131–143. doi: 10.22052/Jns.2018.02.003.

Shi, H., Tang, C., Jianu, A., Fetzer, R., Weisenburger, A., Steinbrueck, M., Mirco Grosse, M., Stieglitz, R. and Müller, G. (2020). Oxidation behavior and microstructure evolution of alumina-forming austenitic and high entropy alloys in steam environment at 1200°C. *Corrosion Science*, 170: 108654. doi: https://doi.org/10.1016/j.corsci.2020.108654.

Shi, S., Zhang, Z. and Yu, L. (2017). Hydrophobic polyaniline/modified SiO2 coatings for anticorrosion protection. *Synthetic Metals*, 233: 94–100. doi: https://doi.org/10.1016/j.synthmet.2017.10.002.

Souto, L.F.C. and Soares, B.G. (2020). Polyaniline/carbon nanotube hybrids modified with ionic liquids as anticorrosive additive in epoxy coatings. *Progress in Organic Coatings*, 143: 105598. doi: https:// doi.org/10.1016/j.porgcoat.2020.105598.

Sumi, V.S., Arunima, S.R., Deepa, M.J., Ameen Sha, M., Riyas, A.H., Meera, M.S., Saji, V.S. and Shibli, S.M.A. (2020). PANI-Fe2O3 composite for enhancement of active life of alkyd resin coating for corrosion protection of steel. *Materials Chemistry and Physics*, 247: 122881. doi: https://doi. org/10.1016/j.matchemphys.2020.122881.

Taheri, N.N., Ramezanzadeh, B., Mahdavian, M. and Bahlakeh, G. (2018). *In-situ* synthesis of Zn doped polyaniline on graphene oxide for inhibition of mild steel corrosion in 3.5 wt.% chloride solution. *Journal of Industrial and Engineering Chemistry*, 63: 322–339. doi: https://doi.org/10.1016/j.jiec.2018.02.033.

Tuken, T., Dudukcu, M., Yazici, B. and Erbil, M. (2004). The use of polyindole for mild steel protection. *Progress in Organic Coatings*, 50(4): 273–282. doi: 10.1016/j.porgcoat.2004.03.004.

Walczak, M., Pineda, F., Fernández, Á.G., Mata-Torres, C. and Escobar, R.A. (2018). Materials corrosion for thermal energy storage systems in concentrated solar power plants. *Renewable and Sustainable Energy Reviews*, 86: 22–44. doi: https://doi.org/10.1016/j.rser.2018.01.010.

Wu, M.J., Lu, L.F., Yu, L.H., Yu, X.Y., Naito, K., Qu, X.W. and Zhang, Q.X. (2020). Preparation and characterization of Epoxy/Alumina nanocomposites. *Journal of Nanoscience and Nanotechnology*, 20(5): 2964–2970. doi: 10.1166/jnn.2020.17460.

Yao, Y.C., Sun, H., Zhang, Y.L. and Yin, Z.D. (2020). Corrosion protection of epoxy coatings containing 2-hydroxyphosphonocarboxylic acid doped polyaniline nanofibers. *Progress in Organic Coatings*, 139. doi: ARTN 10547010.1016/j.porgcoat.2019.105470.

Yeasmin, F., Mallik, A.K., Chisty, A.H., Robel, F.N., Shahruzzaman, M., Haque, P., Rahman, M.M., Hano, N., Takafuji, M. and Ihara, H. (2021). Remarkable enhancement of thermal stability of epoxy resin through the incorporation of mesoporous silica micro-filler. *Heliyon*, 7(1). doi: ARTN e0595910.1016/j.heliyon.2021.e05959.

Yeganeh, M., Nguyen, T.A., Rajendran, S., Kakooei, S. and Li, Y. (2020). Chapter 1—Corrosion protection at the nanoscale: An introduction. pp. 3–7. *In*: Rajendran, S., Nguyen, T.A.N.H., Kakooei, S., Yeganeh, M. and Li, Y. (eds.). *Corrosion Protection at the Nanoscale*. Elsevier.

Yu, Y.-H., Lin, Y.-Y., Lin, C.-H., Chan, C.-C. and Huang, Y.-C. (2014). High-performance polystyrene/graphene-based nanocomposites with excellent anti-corrosion properties. *Polymer Chemistry*, 5(2): 535–550. doi: 10.1039/C3PY00825H.

Zahidah, K.A., Kakooei, S., Ismail, M.C. and Bothi Raja, P. (2017). Halloysite nanotubes as nanocontainer for smart coating application: A review. *Progress in Organic Coatings*, 111: 175–185. doi: https://doi.org/10.1016/j.porgcoat.2017.05.018.

Zhang, D., Gao, L. and Zhou, G. (2006). Molecular design and synergistic effect of morpholinium type volatile corrosion inhibitor. *Journal of Chinese Society for Corrosion and Protection*, 26(2): 120–124.

Zhang, X., Wang, F. and Du, Y. (2007). Effect of nano-sized titanium powder addition on corrosion performance of epoxy coatings. *Surface and Coatings Technology*, 201(16): 7241–7245. doi: https://doi.org/10.1016/j.surfcoat.2007.01.042.

Zheludkevich, M.L., Poznyak, S.K., Rodrigues, L.M., Raps, D., Hack, T., Dick, L.F., Nunes, T. and Ferreira, M.G.S. (2010). Active protection coatings with layered double hydroxide nanocontainers of corrosion inhibitor. *Corrosion Science*, 52(2): 602–611. doi: https://doi.org/10.1016/j.corsci.2009.10.020.

Zhou, C., Hong, M., Yang, Y., Hu, N., Zhou, Z., Zhang, L. and Zhang, Y. (2019). Engineering sulfonated polyaniline molecules on reduced graphene oxide nanosheets for high-performance corrosion protective coatings. *Applied Surface Science*, 484: 663–675. doi: https://doi.org/10.1016/j.apsusc.2019.04.067.

Zhou, C., Lu, X., Xin, Z. and Liu, J. (2013). Corrosion resistance of novel silane-functional polybenzoxazine coating on steel. *Corrosion Science*, 70: 145–151. doi: https://doi.org/10.1016/j.corsci.2013.01.023.

Zhou, Y.N., Li, J.J., Wu, Y.Y. and Luo, Z.H. (2020). Role of external field in polymerization: Mechanism and kinetics. *Chemical Reviews*, 120(5): 2950–3048. doi: 10.1021/acs.chemrev.9b00744.

Chapter 6

Nanomaterials under Biological Conditions

Javier Cortés,[1,4] *Concepción Panadero-Medianero,*[2]
Paola Murgas[3,]* and *Manuel Ahumada*[4,5,]*

1. Introduction

Nanotechnology has been one of the great revolutions in science in the last decades. The translation of working at a bulk level to the nanoscale has prompted several advantages such as higher-performance materials, fine-tuned solutions, and size reduction innovations, among others, that have also given a positive impulse to society by decreasing the cost of materials, allowing the generation of new solutions to current problems, and increasing the access to technology, to name a few. One field that has been particularly favored by nanotechnology is the biomedical area (Bayda et al. 2019). Within this field, currently, it is possible to find several types of nanotechnological solutions like drug delivery, imaging, detection, gene therapy, tissue engineering, and antibacterial, among others, which have been classified as nanomedicine (Lazurko et al. 2019).

Nanomaterials for nanomedicine follow the common descriptions, i.e., they can be defined as materials with sizes between 1 to 100 nm, in at least one of their dimensions, and that have been developed with human intervention. Further, a key

---

[1] Centro de Investigaciones Biomédicas, Escuela de Kinesiología, Facultad de Medicina, Universidad de Valparaíso, Angamos 655, Viña del Mar, Valparaíso.
[2] Center for Integrative Biology, Faculty of Science, Universidad Mayor, Camino La Piramide 5750, Huechuraba, Santiago, RM.
[3] Instituto de Bioquímica y Microbiología, Facultad de Ciencias, Universidad Austral, Valdivia, Región de los Lagos.
[4] Centro de Nanotecnología Aplicada, Facultad de Ciencias, Ingeniería y Tecnología, Universidad Mayor, Camino La Pirámide 5750, Huechuraba, Santiago, RM.
[5] Escuela de Biotecnología, Facultad de Ciencias, Ingeniería y Tecnología, Universidad Mayor, Camino La Pirámide 5750, Huechuraba, Santiago, RM.
* Corresponding authors: paola.murgas@uach.cl; manuel.ahumada@umayor.cl

feature is their high-surface-to-volume ratio, allowing access to nano-engineering the surface by adding/modifying the capping agents/ligands (Ahumada et al. 2019). As with other fields, nanomaterials' compositions for biomedicine can vary depending on the tissue applied or the given biological function. Therefore, carbon, lipidic, polymeric, ceramic, and metallic-based nanomaterials are just a part of the possibilities for developing new solutions (Mitragotri et al. 2015). Figure 1 depicts several types of nanoparticles that have been commonly employed in the literature to treat some organ illnesses.

The development of nanomedicine has allowed relevant advances in treating diseases such as cardiovascular afflictions, Alzheimer's, diabetes, and cancer, just to name a few (Alarcon and Ahumada 2019). Despite these efforts, there is still a gap between society's current health needs and what is done at the laboratory, commonly known as the nanomedicine motto "from the bench to the bedside." This lack of solutions is, in most cases, derived from the resistance developed to current drugs from microorganisms, new viruses, diseases without existing treatment, and regulatory norms (Boucher et al. 2009). Nonetheless, the quick introduction of nanotechnology to medicine has prompted significant improvements in the development of current treatments, e.g., vaccine nano-formulations based on mRNA encapsulation and posterior delivery (Lee et al. 2019, Hofman et al. 2021). As a result of this fast incorporation of nanomaterials into biological environments, the need to understand how these materials interact under biological conditions was relegated to a second plane.

Several researchers and groups have studied how biological systems and vice versa could impact nanotechnology in the last years (Li et al. 2022, Asmatulu et al. 2022, Valdiglesias 2022). Interestingly, in terms of toxicity, most reports highlight

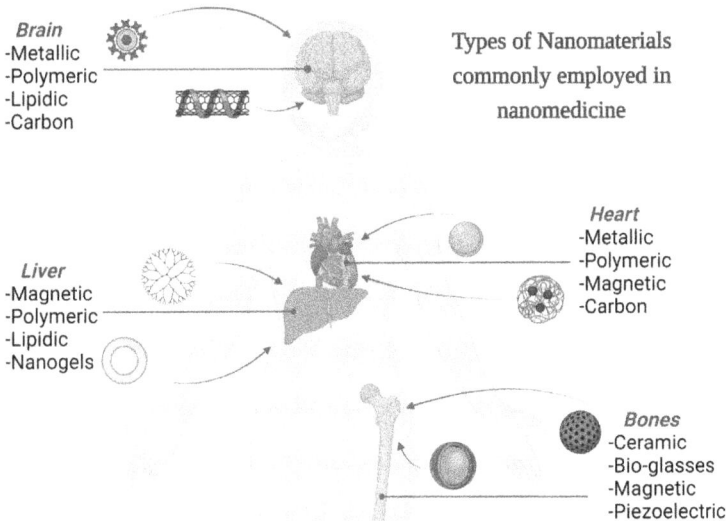

Figure 1. Examples of nanomaterials proposed to be employed in treating specific organ diseases.

the influence of free radicals and their interaction with nanomaterials to impulse the oxidative stress within the biological environment. Therefore, the role of free radicals, their interaction with nanomaterials, and the relevance of oxidative stress are discussed in the following sections.

2. Nanomaterials for Biomedicine

As previously mentioned, nanomaterials intended to be used in medicine follow the same definitions and general considerations that could be applied to any other field using nanotechnology. Therefore, the classification of the nanomaterials employed for nanomedicine can be based on the core composition(s), wherein the literature can be found: carbon-based (nanotubes, fullerenes, graphene, carbon dots), lipid-based (solid lipid nanoparticles and nanostructured lipid carriers), polymer-based (nanocapsules and nanospheres), ceramic-based (oxides, carbides, carbonates, and phosphates of metals or metalloids), and metallic-based materials (Nikzamir et al. 2021). Throughout this chapter, the discussion will focus on metal nanoparticles; nonetheless, the reader should note that most of the topics discussed here could be translated to other types of nanomaterials.

Metal-based nanoparticles (MNPs) had to stand up as one of the predilected nanomaterials intended for biomedicine, founding a site in applications like delivery systems, prophylactic, sensors, hyperthermia, and bioimaging, among others (McNamara and Tofail 2017). The general interest in using MNPs is derived from their physical-chemistry properties. For instance, their synthesis protocols are extendedly sprayed in the literature. Even though they could go through a bottom-up or top-down pathway, the result is usually highly reproducible (Jamkhande et al. 2019). Further, parameters such as size and morphology can be fine-tuned within the same synthesis methodology or *a posteriori* (e.g., by performing new synthesis steps or light irradiation); also, the surface charge can be established by selecting the right capping agent. Other properties such as electric conductivity and the optical phenomena of localized surface plasmon resonance are desirable for treatments that require electric conduction (e.g., for cardiac tissue-related pathologies) and photothermal effect or bioimaging (e.g., theragnostic treatments), respectively (Rai et al. 2016).

Notwithstanding the preceding, toxicity is a specific parameter that must be specially considered for biological systems in contrast with other fields. Thus, significant differences can be found in the capping agent used to stabilize and give function to a nanomaterial for biomedicine. In this regard, common capping agents for biomedicine are DNA, RNA, proteins, antibodies, peptides, amino acids, polymers from natural sources, and/or synthetic polymers innocuous to human health (Javed et al. 2020). As expected, the capping agent will fulfill two leading roles: (i) stabilize the nanomaterial and (ii) give a biological function to the material; examples to clarify this point are offered in Figure 2.

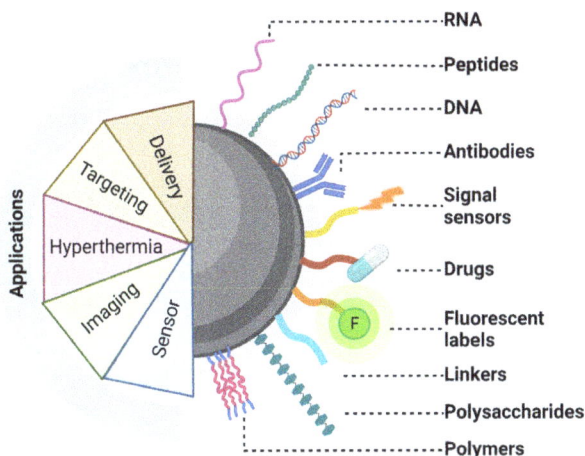

Figure 2. Typical capping agents used in nanomaterials' development and their main applications.

3. Protein Corona

In addition to the selection of the right capping agent to utilize, it is essential to consider that once the nanomaterial is in contact with the biological system, one of the first events to occur is the electrostatic interaction/adsorption of biological molecules with the surface of the material, promoting the formation of a corona effect on the surface (Ahumada et al. 2019). Further, as proteins are those at high presence (concentration) in biological fluids, this effect is commonly referred to as "Protein Corona"; see Figure 3. The corona is a masking process, i.e., if it happens, it will hide the biological function of the nanomaterial and/or modify its stability (i.e., aggregation or precipitation) until clearance from the body or presentation to the immune system (Park 2020). However, it should not be considered static; the corona's formation and persistence on time is a dynamic process that depends on surface potentials, affinity constants, the interaction of additional molecules, and environmental factors such as pH and temperature (Dewald et al. 2015). As depicted in Figure 3, the corona could have several layers covering the nanomaterial, which could vary in composition depending on the parameters above. Undoubtedly, the corona formation usually acts to the detriment of the therapeutic effect of the intended nanomaterial function (Pearson et al. 2014). As so, it should be avoided, particularly in those scenarios where the treatment success depends on the diffusion through the circulatory system. Despite this, a suggestion to circumvent the corona effect is the addition of functional groups that can modify the hydrophilic properties of the MNPs and their half-life and, simultaneously, decrease the interaction with proteins and phagocyte receptors (Dobrovolskaia and McNeil 2007). For example, the use of silencing molecules as capping agents, such as polyethylene glycol (PEG), which binds fewer proteins, and in some cases, it can help to decrease MNPs toxicity; it has been described as a good strategy to increase the therapeutic effect of MNPs (Díaz et al. 2008).

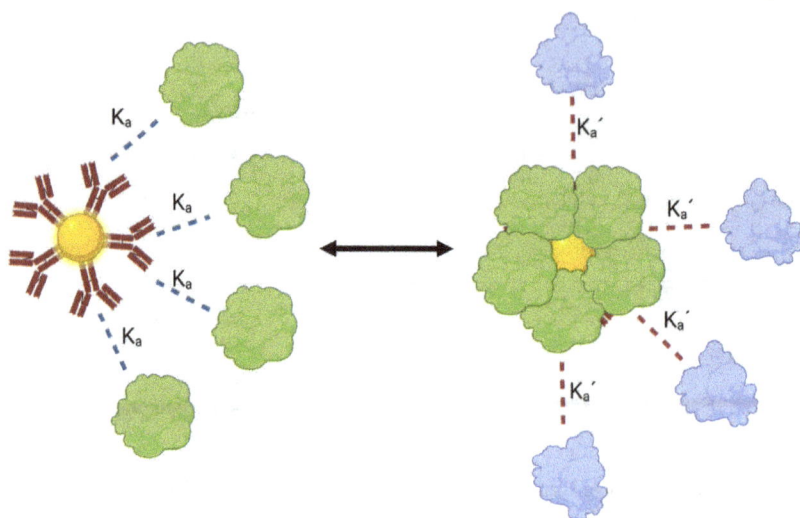

Figure 3. Depiction of the protein corona effect. Left, the nanoparticle treatment faces the presence of the protein. Right, the affinity of the proteins towards the nanoparticle (Ka) masks the therapy, and further layers of proteins can still cover it.

Another relevant consideration related to the protein corona is that human serum albumin (HSA) is the most abundant protein in human blood (around 60% of the total proteins present in the blood); ergo, it is the protein with a higher presence in the circulatory system (Merlot et al. 2014). Therefore, albumin is the protein with higher participation in corona formation and should be considered a good protein model to evaluate a nanomaterial's performance against the corona. Yet, bovine serum albumin (BSA) is commonly used instead of HSA because it is cheaper. While both have high sequence identity percentages (close to 76%) (Huang et al. 2004), the results obtained using BSA should be carefully evaluated, as those are usually interpreted from spectroscopic results, which do not correlate specifically due to an additional tryptophan residue present in the BSA (Steinhardt et al. 1971).

While the rate at which the corona is formed is extremely fast, almost instantaneous in terms of nanoparticle application (Vilanova et al. 2016), if the MNP is well-designed, the protein barrier should be surpassed, and the MNP continues its way to fulfill its therapeutic objective. However, further obstacles will try to block the MNP action, with the immune system being the major barrier.

4. The Immune System

The immune system comprises elements and biological processes that maintain homeostasis against external and internal afflictions. Its function can be divided into two types of responses: innate and adaptive. The innate immune system corresponds to the cells and mechanism that give the organism immediate non-specific defense against foreign elements (Janeway and Medzhitov 2002). In this way, the cells recognize and answer to the exogeneous element in a generic harmful way without establishing a long-term defense memory after the first exposition. On the other hand, the adaptive immune system is organized around two classes of specialized lymphocytes, the T and B cells, which have a diverse repertory of antigen recognition receptors that allow the specific identification and elimination of pathogens. In addition, they oversee the production of immunoglobulins (Ig), i.e., antibodies, that guarantee the immunological memory response (or long-term) against new reinfections (Dunkelberger and Song 2010, Gasteiger and Rudensky 2014). While both types of responses are equally relevant for nanomaterials' characterization, due to their extension as topics and as nanomaterials' first immune challange is the innate response, the role of the adaptive system will not be covered in this chapter; nonetheless, some lectures are recommended to the reader (Schenten and Medzhitov 2011).

4.1 The Innate Immune System and its Cell Components

The immunity cells are produced within the long bones, specifically in their sponged softcore named red bone marrow. Here, the immune cells emerge from the pluripotent hematopoietic stem cells, producing an unlimited number of immune cells (Mahla 2016). Each type of cell comes from a precursor committed to a specific lineage. In this way, red blood cells, platelets, neutrophils, B lymphocytes, and macrophages are produced within the bone marrow to posteriorly flow through the blood and lymphatic circulatory systems (Delves and Roitt 2000, Zlotoff and Bhandoola 2011).

When a foreign body invades the organism, its conserved structures and expressed molecules on its surface or released by the pathogens are recognized as pathogens-associated molecular patterns (PAMPs) (Vance et al. 2009). This recognition is promoted mainly by neutrophils, macrophages, and dendritic cells (Gordon 2002). These cells express patterns recognition receptors (PRRs), which detect the PAMPs present in the membrane of pathogens such as polysaccharides, glycolipids, lipoproteins, and other macromolecules such as nucleotides and nucleic acids (Iwasaki and Medzhitov 2015). The PRRs induce the intracellular activation of a signaling cascade within the innate immune cells, generating the gene activation, which, after transcription, causes a proinflammatory response against the pathogen (Mogensen 2009). Once the pathogen or foreign element is eliminated, these same cells will participate in regenerating the damaged tissue to ensure tissue functionality and homeostasis (Lech and Anders 2013).

The PAMPs detection by PRRs and its related signaling cascade drive the induction of inflammatory responses from the innate cells. These responses are mediated by cytokines, small soluble proteins that act as the communication system between immune cells, and other soluble molecules as chemokines, a subtype of cytokine that facilitate the immune cell migration or recruitment to the target zone that induces the pathogen elimination (Takeuchi and Akira 2010). Furthermore, the pathogen detection by PRRs expressed in antigen-presenting cells, particularly macrophages and dendritic cells, conduce to the activation of the adaptive immune response through the presentation of antigens from the pathogens, which is mediated

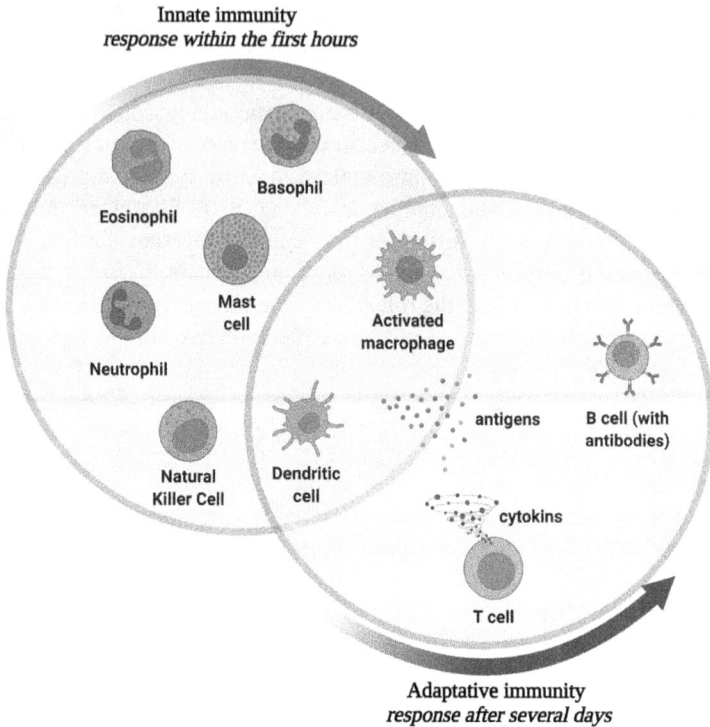

Figure 4. Immune cells from mammals. The immune system consists of innate (fast) and adaptative (slow response) immunity. In the first type of immunity, basophil, eosinophil, mast cell, neutrophil, natural killer, macrophages, and dendritic cells can be found. The last two types of cells are also relevant in inducing the adaptative immunity by presenting the antigens. The adaptative system involves the B and T cells.

by the major histocompatibility complex II (MHCII). Once the antigens are exposed to the MHCII, these are recognized by the T cells, allowing the activation and communication between the innate and adaptative systems (Janeway 1989), see Figure 4.

5. Macrophages

Considering the relevance of each type of immune cell, macrophages stood up as one of the most exciting cells in the flow pathway of nanomaterials; therefore, this section focus on its main characteristics and functions.

Macrophages are cells that arise from the bone marrow as monoblasts, which *a posteriori* differentiate into promonocytes and monocytes, establishing the mononuclear phagocyte system. Once the monocytes migrate from the marrow, they circulate through the blood or lymph until they finally are recruited into different tissues, where they mature and become macrophages. Further, macrophages can move to lymphatic organs, like the spleen, to present antigens to the T lymphocytes (Das et al. 2015). In terms of response time, macrophages are the immune cell type that arrives in second place, after neutrophils, to the damaged zone or infection (neutrophils are the first ones); such migration occurs within minutes and can last for hours (Prame Kumar et al. 2018).

For macrophages to reach their triggering, it requires going from a resting state to an activated state. Likewise, several signals are needed; among them, the interferon-gamma cytokine (IFN-γ), which binds to its receptor expressed onto the macrophage's plasmatic membrane, inducing a cellular signaling cascade that, in addition to a second signal derivate from the recognition of a PAMP, induces the activation of the macrophage by acquiring a phenotype with specific functions (Martinez and Gordon 2014). The interaction with IFN-γ and/or PAMPs promotes the macrophages to reach a type 1 phenotype (M1), commonly known as the proinflammatory phenotype, secreting the Tumor Necrosis Factor-alpha (TNF-α), Interleukin 1 beta (IL-1β), Interleukin 12 (IL-12), all molecules that facilitated, among other, the generation of T helpers 1 lymphocytes (Th1), synthesis and secretion of reactive oxygen and nitrogen species (ROS and NOS, respectively, see Figure 5). These last ones facilitate the elimination of pathogens, tumoral cells, and foreign elements that represent potential damage to the organism (Martinez and Gordon 2014).

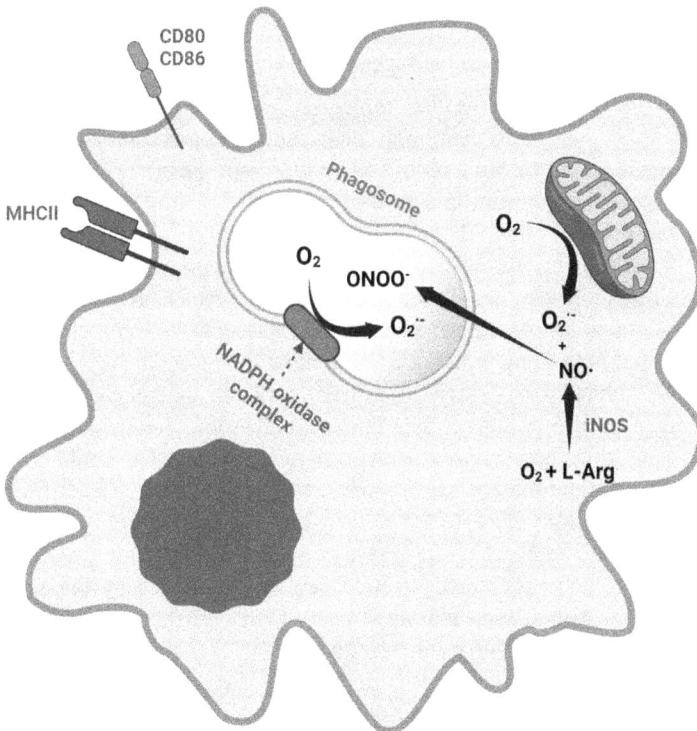

Figure 5. ROS generation mechanism by macrophages. The electron transport chain produces the superoxide radical anion ($O2^{\cdot-}$) at mitochondrial level and by the NADPH oxidase complex. Nitric oxide ($NO^{\cdot}$) can be generated by activating the nitric oxide synthase (iNOS) by arginine. Both types of radicals can produce peroxynitrite ($ONOO^{-}$). CD80, CD86, and the Major Histocompatibility Complex II (MHCII) are molecules involved in the antigen presentation and are considered as a marker for macrophage activation.

Macrophages can also acquire an anti-inflammatory phenotype (M2). Reaching such a phenotype requires binding Interleukin 4 (IL-4) or Interleukin 13 (IL-13) with its respective membrane receptor on the macrophage. M2 macrophages act by secreting anti-inflammatory interleukins such as the Interleukin 10 (IL-10), Transforming Growth Factor-beta (TGF-β), and Insulin-like Growth Factor 1 (IGF-1) that, finally, promote the tissue reparation and regeneration, and angiogenesis (Martinez and Gordon 2014).

For macrophages to recognize foreign elements, it requires the PRRs where to date, four prominent families have been identified (Gustafson et al. 2015).

i) Scavenger receptors: Integral membrane proteins that bind different types of polyanionic ligands with high affinity (Suzuki et al. 1997).

ii) Toll-like receptors (TLRs): Integral transmembrane glycoproteins expressed in the plasmatic membrane and membranes of intracellular vesicles (such as endosomes and lysosomes). They oversee the recognition of PAMPs and damage-associated molecular patterns (DAMPs) (Takeda et al. 2003).

iii) Fc receptors allow cells to recognize the antibodies bound to the microorganism and infected cells' surfaces (Swanson and Hoppe 2004).

iv) Mannose receptor: A transmembrane protein that binds to the carbohydrates presented on the surface of pathogens (Mogensen 2009).

In the first instance, the macrophages generate tentacles-like extensions derivate from the plasmatic membrane or pseudopods and lamellipodia that wrap the exogenous element to carry on the recognition and posterior degradation of the pathogens into the cytoplasm and further promote the formation of a phagocytic vesicle (phagosome). Then, the phagosome is merged with vesicles containing proteases at high pH (either acid or basic), i.e., lysosome, and from here, there are denominated phagolysosomes (Gordon 2016). The mechanisms required to destroy the foreign element are developed inside this last vesicle (Gagnon et al. 2002). Part of these mechanisms contemplates ROS generation through the NOX enzymatic complex (Schumann 2016), which can degrade collagen, inactivate enzymes, oxidate lipids, damage the DNA, attack membranes, and promote cellular lysis (Wang et al. 2017). Note that these consequences are relevant, as ROS could alternate the phagosome membrane integrity and consequently allow the release of the ROS to the cytoplasm, and from there, to the cell's exterior, generating collateral damage in neighbor's cells and adjacent tissue (Erard et al. 2018). Fortunately, under balanced conditions, the toxicity promoted by ROS can be neutralized by the action of the antioxidant mechanisms, for example, enzymes such as the superoxide dismutase (SOD) at the cellular and tissular level (Indo et al. 2015). To conclude the phagocytic pathway, once the external element is digested, it stays in a vesicle containing the garbage or the antigen (if this could not be degraded), then by exocytosis and in contact with the MHCII, they are presented to the T lymphocytes involved in the adaptive immune response (Lim et al. 2017).

6. Free Radicals, Reactive Oxygen Species, and Oxidative Stress

Previously, it was mentioned that ROS has a vital role in eliminating foreign elements in the organism. This section describes their main characteristics as their consequences under oxidative imbalance conditions.

Free radicals are molecules that have an unpaired electron in their external orbital, giving them a high instability configuration (Redza-Dutordoir and Averill-Bates 2016). At the biological level, they are formed as byproducts of the normal oxygen metabolism (Mailloux and Harper 2011), so the oxygen (O_2), after internal chemical reactions, enzymatic actions, or environmental factors, becomes able to transform into a free radical, namely reactive oxygen species (ROS), with the potential to damage other elements (Nosaka and Nosaka 2017). Among the most known ROS are the superoxide anion ($O_2^{\cdot-}$), hydrogen peroxide (H_2O_2), singlet oxygen ($^1O_2$), and hydroxyl radical ($^{\cdot}OH$) (Mailloux 2015). They have a short half-life (milliseconds), and their reactivities are inversely correlated to such a half-life (Hrycay and Bandiera 2015).

ROS are essential modulators for cellular functions. At low concentrations, it has been described that they participate in the cellular signaling, mitogenic response induction, and defense against infectious agents, among others (Buonocore et al. 2010). For example, H_2O_2 act as a molecular messenger through the oxidative modification of signaling proteins (Widlansky and Gutterman 2011). Therefore, the balanced conditions between ROS production and their remotion allow a proper cell function (Halliwell 2012). However, if the ROS production overpasses the remotion mechanisms within the biological tissues, an oxidative imbalance is reached, commonly known as oxidative stress. This stress can be defined as the exposure of a given element to diverse sources that produce an equilibrium disruption between prooxidant substances or factors and antioxidant mechanisms (Sies et al. 2017). Under the context of the biological environment, the oxidative stress leads to a series of reactions that produce oxidative damage within different types of biological molecules, such as lipids, proteins, and DNA, to mention a few (see Figure 6), deteriorating and inactivating the cellular functions (Pisoschi and Pop 2015).

7. Metal Nanoparticles and the Oxidative Stress

MNPs in biomedicine have been extensive; nonetheless, they have been described as ROS promotors inducing toxicity and metabolism interference in living organisms (Tang et al. 2016). Note that ROS sources can also come from the exterior, for example, because of the metabolism of some substances (chloroform, paracetamol, ethanol, etc.), or in a direct manner because of the presence of transition metals (iron, copper, silver, gold, etc.) (Flora et al. 2008). In the presence of reduced transition metals, the H_2O_2 is partially reduced, generating the $^{\cdot}OH$, one of the most potent oxidants in nature, in a process known as the Fenton reaction (Halliwell 2013).

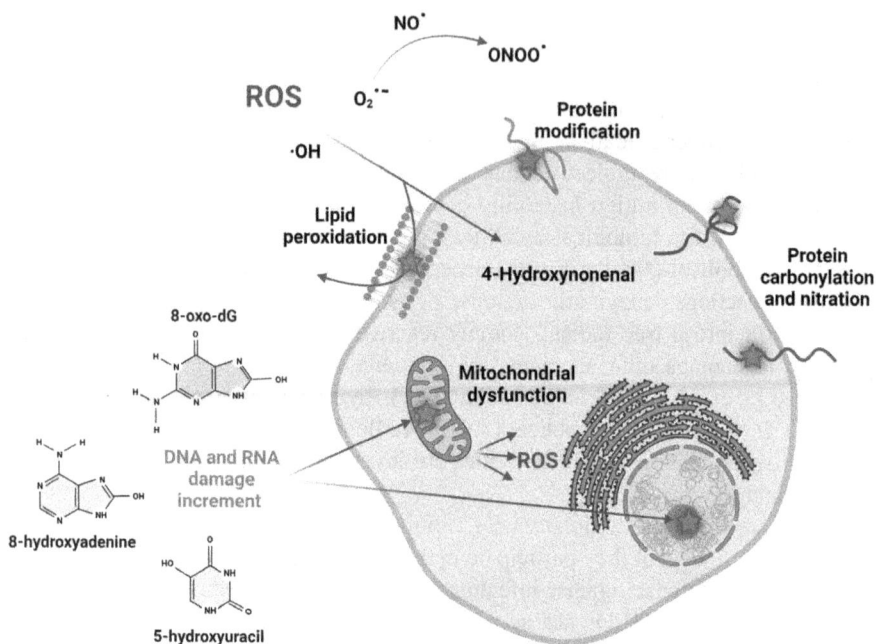

Figure 6. The oxidative stress and its potential damage to the cell components. Oxidative stress is associated with high levels of oxidation products such as proteins, lipids, and nucleic acids, promoted by the reactive oxygen species (ROS). The red stars indicate oxidative damage.

Additionally, in living systems, the presence of H_2O_2 and $O_2^{\cdot-}$ can generate more $^{\cdot}$OH, in a process known as the Haber-Weiss reaction (Kehrer 2000). This last free radical reacts so fast with biomolecules that no antioxidant, at physiological concentrations, can compete with the $^{\cdot}$OH generated to protect the molecules. Then, the best strategy to minimize the oxidative damage by $^{\cdot}$OH is to remove the H_2O_2 or chelate the transition metals (Halliwell 2013). Under both scenarios, proteins such as transferrin, ceruloplasmin, and albumin must act as extracellular chelators of the metal ions, avoiding the toxicity of the hydroxyl radical (Biswas 2016). Well-known is that in aqueous dissolutions, reduced metals can be oxidized to their respective metal ion (Behra et al. 2013). This is how the debate about the potential toxicity of MNPs and/or their released ions to the environment can promote over cells (Barros et al. 2018). In this way, those toxicity mechanisms are not entirely established. Still, there is a high consensus that their toxicity can be associated with direct contact of the MNPs with the cellular membranes, followed by their internment in the cellular cytoplasm (Gahlawat et al. 2016).

Some of the most studied MNPs for ROS generation are the silver nanoparticles (AgNPs), which are tightly related to their antimicrobial activity. This property shares with other MNPs such as gold, copper, and zinc nanoparticles. Besides the fact that some mechanistic routes are still unknown, the recollected information on ROS and AgNPs has allowed a general view of the free radical production because

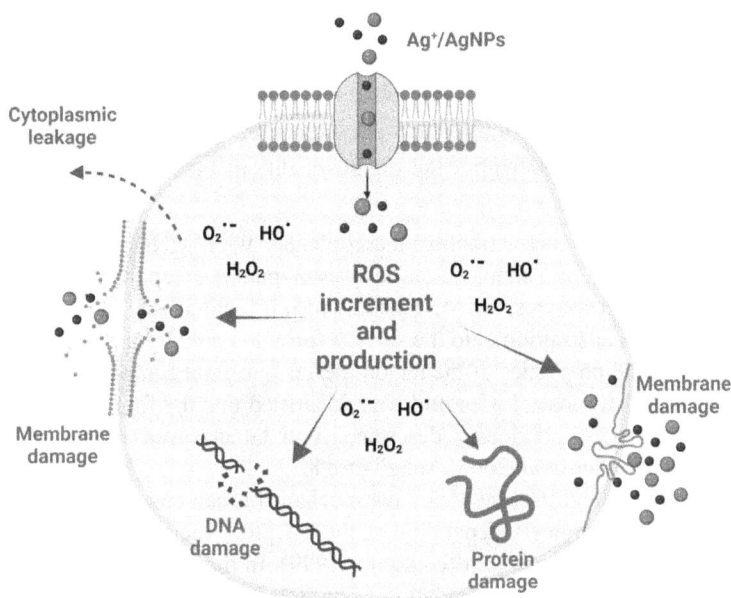

Figure 7. Proposed antimicrobial mechanism of silver ions and/or AgNPs. In a general manner, in the first instance, the AgNP and/or silver ions (Ag^+) can bind or permeate the cellular membrane, posteriorly it penetrates the cells, to finally produce oxidative damage by ROS generation in the DNA and proteins, to name a couple of oxidative targets.

of MNPs, see Figure 7. These mechanisms can be summarized in three main events (Le Ouay and Stellacci 2015, Salem et al. 2015, Biswas 2016):

i) Due to the controlled release of metal ions (silver in this case), the production of adenosine triphosphate (ATP) and DNA replication are inhibited, both fundamental for cell survival.

ii) MNPs can directly promote damage to the cell's membrane and biomolecules (lipids, proteins, and DNA).

iii) MNPs can generate ROS, specially ˙OH, promoting cellular death.

8. MNPs and the Immune System

Previously, the role of the immune system in the defense against foreign elements was described; in this section, the MNPs as "foreign elements" and their interaction with the immune system are discussed.

MNPs can be recognized and interact with the immune system and modulate its function by inducing an immunosuppressor or immunostimulatory effect (Dobrovolskaia 2015). The chemical and structural diversity allows the MNPs to reach different cells and tissues, making them useful for biomedicine. However, these characteristics and properties promote the MNPs to be recognized when entering the organism (Dobrovolskaia et al. 2016). This recognition would depend upon factors such as doses, administration route, size, composition, and surface properties of the

MNP (Reichel et al. 2019), whereas the immune cells can be exposed to the MNPs in three main ways (Walkey et al. 2012):

i) Oral or parental administration of pharmaceutical formulations based on MNPs.
ii) MNPs' inhalation by pollution or work exposure.
iii) Generation of MNPs within the organism due to the degradation of metallic implants.

Considering the interaction of macrophages with MNPs, since exogenous elements, pathogens, and damaged tissue, present patterns recognized by the surface receptors of the phagocytes, the MNPs could also present analog molecular patterns due to the protein adsorption onto the surface (protein corona), or due to the specific physical-chemical properties of the nanomaterial's original composition (Gustafson et al. 2015). In this sense, Taylor and cols. identified that the four surface receptors of the macrophages, previously described, will be involved in recognizing the nanomaterials (Taylor et al. 2005), see Figure 8.

Further, the internalization of cellular mechanisms can be generally classified as phagocytosis or pinocytosis, depending on the receptors, membrane mechanism, and cellular disposition (Aderem and Underhill 1999). In the case of phagocytosis, it is used for the absorption and degradation of particles of sizes over 500 nm (Doherty and McMahon 2009). Its phagocytosis rate varies widely and changes depending on the type of cell, activation state, cultivation condition, or biological conditions in which the MNPs are found (e.g., exposed to endotoxins or proteins) (Kumari et al. 2010).

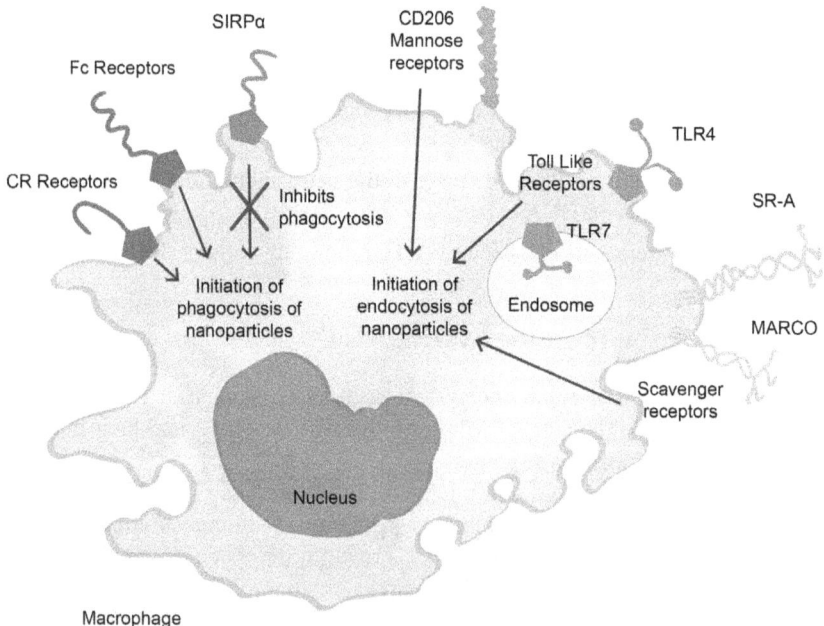

Figure 8. Macrophage's surface receptors and the recognition of nanomaterials. Each receptor that recognizes an MNP will induce a specific internalization mechanism, as depicted. Posterior inflammatory effects, triggered by receptor participation, can be caused by these internalization pathways.

On the other hand, pinocytosis is activated in every mammalian cell and is responsible for the MNPs absorption. This caption mechanism can be classified as significant volume extracellular internalization (e.g., macropinocytosis) and a small volume internalization, for example, in a base of clathrins, able to internalize particles with sizes ranging from 20 to 500 nm (Doherty and McMahon 2009, Kumari et al. 2010). The scavenger, TLRs, and mannose receptors participate in these mechanisms of recognition and induction, which involve the formation of clathrin invaginations in the cellular membrane that wrap and drag the cargo to the cell's interior, in a process called phagocytosis (McMahon and Boucrot 2011). Then, these membrane invaginations, through the scission by dynamin, form endocellular vesicles that are sent to early endosomes, acidified, and transformed into late endosomes, that are further directed to other intracellular destinies, including the lysosome's compartments (Doherty and McMahon 2009, Canton and Battaglia 2012), see Figure 9.

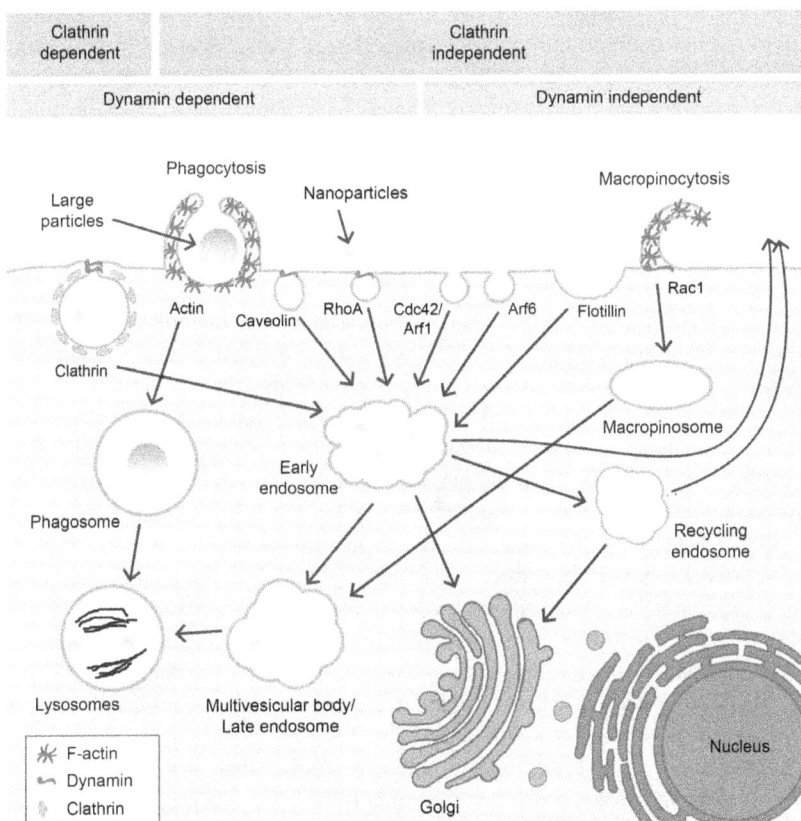

Mechanisms of endocytosis:

Clathrin dependent	Clathrin independent
Dynamin dependent	Dynamin independent

Figure 9. Endocytic mechanisms and intracellular transport model. Nanoparticles (light grey dots) and other captured substances by endocytosis are locked up within early endosomes, phagosomes or macropinosomes. Posteriorly, these vesicles mature, by a degradative pathway, becoming multivesicular bodies (late endosomes) that fuse with lysosomes. Alternatively, the MNPs can be transported to the cellular surface directly from the early endosomes or by the recycling endosomes.

9. Oxidative Stress: Macrophage's Toxicity Derived from the MNPs Action

The ROS responses mediated by MNPs orchestrate a series of pathological events such as genotoxicity, inflammation, fibrosis, and carcinogenesis (Li et al. 2010a). The specific mechanisms that reign the ROS production due to the MNPs' presence have not been fully elucidated to date. Despite this, the most relevant mechanisms proposed in the literature are presented.

9.1 Toxicity Mediated by Transition Metals

Most metallic nanomaterials promote toxicity mediated by free radicals through the Fenton reactions (Abdal Dayem et al. 2017). It has been demonstrated that MNPs such as titanium dioxide (TiO_2), zinc oxide (ZnO), and AgNP, among others, mainly act by depositing on the cell surface or inside the cellular organelles inducing oxidative stress signaling cascade (Gustafson et al. 2015). CuNPs follow a similar pathway; their antimicrobial activity is derived from the copper ions released to the medium that will further damage the macrophage cell membrane, altering its function and driving it to death (Ren et al. 2009, Sohrabnezhad et al. 2014). In this line, studies of AuNP internalized in human macrophages have shown an increase in the ROS intracellular production (Fatima et al. 2015).

9.2 Immune System Activation and Inflammation Mediated by MNPs

MNPs recognition by the immune system activates the inflammatory cells such as neutrophils and macrophages through their influence on the intracellular calcium concentrations, activation of transcriptional factors, and modulation of the cytokines production, resulting in an increment in the production of ROS, and by consequence generating oxidative stress (Manke et al. 2013). This last one unleashes cellular signaling pathways that, in the end, increase the expression of the proinflammatories and fibrotic cytokines (Li et al. 2010a).

Further, it has been demonstrated that AuNP can initiate an inflammatory response mediated by macrophage activation both *in vitro* and *in vivo* (Fallarini et al. 2013). Another study compared the exposition of macrophages to AuNP and AgNPs, concluding that the short-term exposition to AgNP induced the signaling of the nuclear factor kappa B (NF-κB), considered the main transcriptional factor in the inflammatory response, and ROS production, driving an increase in the levels of proinflammatories cytokines such as TNF-α and IL-6 (Nishanth et al. 2011). In 2010, Yazdi and cols. observed similar results after applying TiO_2 (20 and 80 nm), generating ROS production. They established that the MNPs were responsible for the inflammasome activation, multiprotein complexes that promote the caspase-1 activation, through the synthesis and secretion of IL-1β in macrophages. The adapted protein MyD88, part of the TLRs signaling pathway responsible for the translocation downstream of the NF-κB factor to the nucleus, also participates in this process. The activated signaling pathway by the bonding of IL-1β to its receptor additionally activates the caspase-1 protease, which oversees the inflammatory processes such as apoptosis, generating, in an ultimate instance, a constant proinflammatory vicious circle (Yazdi et al. 2010).

9.3 Mitochondrial Disruption Mediated by MNPs

Apoptosis is considered one of the main cellular death mechanisms caused by oxidative stress induced by MNPs. Among the different apoptotic pathways, the mitochondrial apoptotic pathway plays a relevant role in cellular death, as these are one of the prominent organelles' targets of the MNPs (Ma and Yang 2016). First, the MNPs damage the phospholipids from the mitochondrial membrane, provoking its depolarization. Then, the MNPs enter the mitochondria and overstimulate the ROS production by disrupting the electron transport chain (Manke et al. 2013). The released metallic ions interact with the available reduced thiol groups from the enzymes and proteins, interfering with the respiratory chain, affecting the mitochondrial ATP production, altering the membrane permeability, morphology, structure, and function of this organelle key for the metabolism and cell survival (Hsueh et al. 2015). Another consequence is chromosomic DNA alteration, which affects the replication capacity and the synthesis of essential proteins. Several MNPs, including Zn, Cu, Ag, and Fe-derived nanoparticles, provoke the cellular death mediated by ROS through the mitochondrial disruption, exhibiting high levels of caspase-3 and caspase-9, and high proportions of protein 4 similar to bcl-2 (Bax)/B-2 cells lymphoma (Bcl-2) (Abdal Dayem et al. 2017, Liao et al. 2019, Chernousova and Epple 2013).

9.4 Phagosome Instability and Lysosome Generation by MNPs

It has been proposed that there exists a synergic effect in the interaction between phagocytes and MNPs. On one side, the phagosome activation, potentiated by the cellular stress, promotes an increase in the ROS release and cytokine production (IL-6, IL-12, IL-1β, TNF-α, INF-γ). On the other, the direct contact of the MNPs with significant portions of the phagocyte's membrane surface causes damage and integrity lost to this last one, triggering the cellular content leak and consequently affecting the cell's survival (Barros et al. 2018, Bondarenko et al. 2013).

Previously it was mentioned that after the MNP's phagocytizing, they end up in endocytic vesicles that finally merge with the lysosomes, an organelle with a low pH due to the presence, in its membrane, of a proton pump ATPase that transports protons (H^+) from the cytosol to the interior (Xia et al. 2008). This facilitates the proteases' function and hydrolytic enzymes, which oversee the degradation of the biological substance (Mindell 2012). In 2011, researchers demonstrated the peptide capped AuNP (16 nm; positive charges) caption by endosomes. Once merged with the lysosome, it was observed that the MNPs could absorb protons in response to the endosome acidification, thus acting as "proton sponges" (Iversen et al. 2011). The MNPs could use this to disrupt the membranes of the vesicles and release proteases and enzymes to the cytosol, promoting cytotoxicity and inducing cell death's signaling cascades (Sharifi et al. 2012).

9.5 Macromolecules Damaged by MNPs

As established before, ROS are particularly dangerous for lipids, proteins, and nucleic acids. The DNA is one of oxidative stress's primary target and represents the first step involved in the mutagenesis, carcinogenesis, and cellular aging (Shi et al. 2004). The genotoxic effects can be produced by direct interaction of the MNPs with genetic material or by collateral damage promoted by ROS, previously caused by the MNPs. In this sense, the MNPs can induce chromosomic aberrations, DNA chain breaks, DNA oxidative damage, and mutations (Xie et al. 2011). Particularly, it is known that the ·OH reacts with all DNA's constituents promoting its breakage into a single strand due to the adduct 8-hydroxy-2′-deoxyguanosine (8-OHdG) formation, which is a DNA oxidative damage biomarker (Valavanidis et al. 2009). In this sense, by using this biomarker, in 2011, a study compared the genotoxicity promoted by ROS when using different MNPs (Cu, Fe, Ti, and Ag); in all the cases, the results showed the formation of micronucleus and DNA oxidative damage *in vivo* (Song et al. 2012).

ROS can also interact with lipids, which are abundant in the biomembranes, to produce lipidic peroxidation byproducts associated with the mutagenesis. During the initial steps of ROS generation, the polyunsaturated fatty acids are prone to oxidation to form lipidic hydroperoxides (Ayala et al. 2014). Then, the prooxidant metals such as Cu and Fe react with them to induce final products, with the ability to damage the DNA, like the malondialdehyde (MDA) and the 4-hydroxynonenal (4-HNE), both acting as mediators for inflammatory response and as carcinogenesis risk factors (Napierska et al. 2012, Shukla et al. 2011, Turski and Thiele 2009). Thus, the free radicals induced by MNPs, or their metal ions, can activate oncogenes like GTPase activating protein (RAS), where it has been proved that excessive amounts of MNPs are associated with diverse types of cancers among them: skin, bladder, liver, lung, and respiratory tract (Manke et al. 2013).

Like lipids, proteins are also an easy target for ROS. Among the main modifications induced in proteins by oxidation are catalytic activity loss, aminoacidic modifications, carbonyl group formation, thermal stability alteration, viscosity changes, fragmentation, intra- and inter-proteins crosslink, disulfide bridge formation, and a higher tendency to proteolysis. It must be considered that amino acids are prone to oxidation (such as proline, lysine, and arginine); nonetheless, a protein, depending on its tridimensional conformation, can or cannot expose those amino acids to oxidation (Nyström 2005). Despite this, the ROS interaction with proteins promotes carbonylation on the protein's lateral chains, among them, carbonyl groups (aldehydes and ketones) formation, histidine residues oxidation to oxo-histidine, and other degradation byproducts such as those obtained in the reaction of phenylalanine to Orto- and Meta-tyrosine, methionine conversion to methionine sulfoxide or the oxidative degradation of tryptophan to kynurenines (Nyström 2005, Thanan et al. 2014). In 2010, a study that evaluated the autophagy and oxidative stress promoted by AuNPs established that the exposition to these MNPs double-fold the protein oxidation measured through the formation of the MDA-protein adduct compared to the control (Li et al. 2010b). Another research performed a proteome analysis of macrophages proteins when exposed to 11 types of MNPs, finding that all of these can promote protein translocation and unfolding because of the ROS overproduction (Zhang et al. 2020). Other work evaluated the effect of cobalt and chromium ions derived from prostheses in macrophages; the authors found that metal ions effectively promoted ROS production and protein oxidation. Nonetheless, such findings were observable only in the cytoplasm and not in the nucleus (Petit et al. 2005).

9.6 Toxicity Derived from the MNP's Size and Surface Area

It has been observed that there is a direct correlation between MNP's smaller sizes and toxicity (Ivask et al. 2014). In 2008, a study demonstrated that when comparing AgNPs of different sizes (5 and 11 nm), the smaller ones could release larger quantities of the Ag^+ ion to the medium (Jung et al. 2008). Other reports have also shown that the surface area stimulated ROS production. For instance, a study that compared the cytotoxicity of AgNPs of different sizes (15 and 50 nm) but same concentrations established that the smaller AgNPs induced higher-level production of ROS in macrophages (Cho et al. 2009).

Summarizing, immune system cells can recognize the MNPs as foreign elements. Because of that, they promote a series of free radicals and inflammatory responses to eliminate them (Figure 10). At the same time, the inflammation can

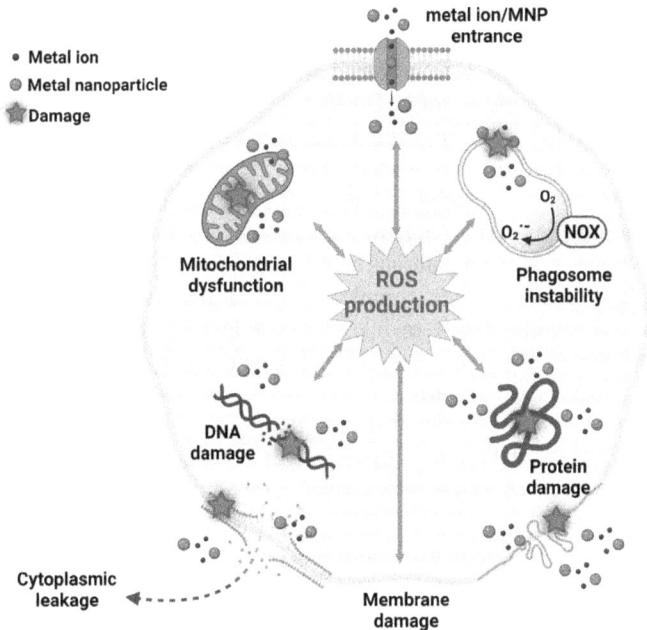

Figure 10. Proposed phagocyte toxicity mechanism when exposed to oxidative stress induced by MNPs. The MNPs enter the cell through phagocytosis, integral membrane proteins, or endocytosis. Once inside, metal ions are released, prompting ROS production. The ROS excess drive oxidative damage in the mitochondrial membrane and electron transport chain, promoting its dysfunction and apoptosis. After interacting with the MNPs, the phagosomes generate high amounts of ROS through the NOX enzymatic complex, causing damage to the phagosome membrane and later leakage to the cytoplasm. ROS overproduction can also prompt DNA damage, chain rupture, protein denaturalization, and lipid peroxidation from oxidative stress.

potentiate ROS production. The action mechanism that shares the MNPs is the interaction they perform over the target cells' external membranes, leading to membrane disruption, permeability modifications, cellular respiration alterations, and finally, cell death (Hajipour et al. 2012). The oxidative stress related to MNP's exposition implies mitochondrial respiratory chain dysfunction, mitochondrial apoptosis, NADPH oxidase system activation, macromolecules oxidative damage, and natural antioxidant enzyme depletion; all of them are associated with tissular damage (Eftekhari et al. 2018). The ROS generated because of the MNPs contributes to the activation of the cellular signaling, inflammatory cytokines and chemokines expression, and specific transcription factors activation such as NF-κB (Manke et al. 2013). The activation of the cellular mechanism is tightly related to the transcription of the gene involved in inflammation, genotoxicity, and cancer development. Therefore, the pathologic consequences observed during the exposition of biological systems to MNPs can be attributed to ROS generation (Abdal Dayem et al. 2017).

10. Outlook and Future Perspectives

Nanomedicine has played a pivotal role in advancing biomedicine by tackling several obstacles that current clinical treatments face. Nanomaterials can be designed and fine-tuned from scratch, i.e., parameters such as composition, shape, size, surface charge, and capping agents, among others, are adjustable through the chosen synthetic pathways to fulfill specific biological needs like delivery, imaging, sensors, targeting, and hyperthermia. In this sense, nanomaterials have proven to be a valuable tool for society. More expected formulations and applications are on their way to coming to light, as others are in the process of approval for organisms such as the U.S. Food and Drug Administration (FDA). However, the success should not blind us about the potential toxicities that nanomaterials could promote in human health.

Biological organisms are complex systems that have evolved to protect themselves and facilitate their own repairment in case of damage. Most of this work is overseen by the immune system. The immune system is divided into two subsystems: innate and adaptative, where macrophages have a central role in the leading response to foreign elements, degradation, and posterior presentation to the adaptative immune system. All of this is under a normal function of the biological machinery. However, under unstable oxidative conditions, oxidative stress prompts several proinflammatory cellular signaling cascades, membrane disruption, DNA and protein damage, and other afflictions that potentially lead to cell death and tissue damage.

The current use of nanomaterials, particularly metal nanoparticles, has focused on the therapeutic effect, but lower attention has been given to their stability and potential toxicity. Once in contact with a biological organism, MNPs interact rapidly with proteins to form the protein corona, which can mask the therapeutic effect of the nanomaterials and/or promotes its instability. Further, the immune system will recognize them as exogenous elements and be treated according to them. While macrophages are not the first immune cells to reach them, they are in a prolonged period and are also in charge of presenting the MNPs to the adaptive immune system. Moreover, the interaction of MNPs and macrophages provoke a vicious cycle of oxidative stress that can affect other cells and tissues.

This chapter covered only the essentials related to the interaction of nanomaterials with biological organisms. It should be considered that this is only the tip of the iceberg in this matter. Nanomedicine is the future of biomedicine, without a doubt. Still, as seen in this chapter, biological systems are a clear example of an extreme condition where several parameters should be considered for the nanomaterial preparation and its posterior evaluation.

Future works should continue exploring the wonders that nanotechnology can bring to biomedicine to help society have better life qualities. Nonetheless, one must not forget that nanomaterials' stability and potential toxicity are essential to their therapeutic effect. In this sense, the active role of capping agents as the first line of interaction with the biological systems should be further explored, particularly the possibility of increasing their pivotal role in preventing the oxidation of MNPs. Further, the interaction of MNPs with the immune system should be deeply explored at the innate and adaptive levels. Finally, oxidative stress is a clear marker of damage and aging; therefore, a new quest to avoid and fight against it through nanotechnology is a topic with great potential for humanity.

Acknowledgments

MA thanks ANID-FONDECYT for grant project 11180616 and Universidad Mayor for its support. PM thanks ANID-FONDECYT for grant project 11190258. Figures were created with BioRender.com.

References

Abdal Dayem, A., Hossain, M.K., Lee, S.B., Kim, K., Saha, S.K., Yang, G.M., Choi, H.Y. and Cho, S.G. (2017). The Role of Reactive Oxygen Species (ROS) in the biological activities of metallic nanoparticles. *Int. J. Mol. Sci.*, 18.

Aderem, Alan and David M. Underhill. (1999). Mechanisms of phagocytosis in macrophages. *Annual Review of Immunology*, 17: 593–623.

Ahumada, Manuel, Caitlin Lazurko and Emilio I. Alarcon. (2019). Chapter 1—Fundamental Concepts on Surface Chemistry of Nanomaterials. *In*: Julia Pérez Prieto and María González Béjar (eds.). *Photoactive Inorganic Nanoparticles* (Elsevier).

Alarcon, Emilio I. and Manuel Ahumada. (2019). *Nanoengineering Materials for Biomedical Uses* (Springer Nature: Switzerland).

Asmatulu, Eylem, Mohammad Nahid Andalib, Balakrishnan Subeshan and Farhana Abedin. (2022). Impact of nanomaterials on human health: A review. *Environmental Chemistry Letters*, 20: 2509–2529.

Ayala, Antonio, Mario F. Muñoz and Sandro Argüelles. (2014). Lipid peroxidation: Production, metabolism, and signaling mechanisms of malondialdehyde and 4-hydroxy-2-nonenal. *Oxidative Medicine and Cellular Longevity*, 2014: 360438–38.

Barros, Caio H.N., Stephanie Fulaz, Danijela Stanisic and Ljubica Tasic. (2018). Biogenic Nanosilver against Multidrug-Resistant Bacteria (MDRB). *Antibiotics (Basel, Switzerland)*, 7: 69.

Bayda, Samer, Muhammad Adeel, Tiziano Tuccinardi, Marco Cordani and Flavio Rizzolio. (2019). The history of nanoscience and nanotechnology: From chemical-physical applications to nanomedicine. *Molecules (Basel, Switzerland)*, 25: 112.

Behra, R., Sigg, L., Clift, M.J., Herzog, F., Minghetti, M., Johnston, B., Petri-Fink, A. and Rothen-Rutishauser, B. (2013). Bioavailability of silver nanoparticles and ions: From a chemical and biochemical perspective. *J. R. Soc. Interface.*, 10: 20130396.

Biswas, S.K. (2016). Does the interdependence between oxidative stress and inflammation explain the antioxidant paradox? *Oxid. Med. Cell Longev.*, 2016: 5698931.

Bondarenko, Olesja, Angela Ivask, Aleksandr Käkinen, Imbi Kurvet and Anne Kahru. (2013). Particle-cell contact enhances antibacterial activity of silver nanoparticles. *PLoS ONE*, 8: e64060.

Boucher, Helen W., George H. Talbot, John S. Bradley, John E. Edwards, David Gilbert, Louis B. Rice, Michael Scheld, Brad Spellberg and John Bartlett. (2009). Bad bugs, no drugs: No ESKAPE! An update from the Infectious Diseases Society of America. *Clinical Infectious Diseases*, 48: 1–12.

Buonocore, G., Perrone, S. and Tataranno, M.L. (2010). Oxygen toxicity: Chemistry and biology of reactive oxygen species. *Semin. Fetal. Neonatal. Med.*, 15: 186–90.

Canton, Irene and Giuseppe Battaglia. (2012). Endocytosis at the nanoscale. *Chemical Society Reviews*, 41: 2718–39.

Charles A. Janeway, Jr. and Ruslan Medzhitov. (2002). Innate immune recognition. *Annual Review of Immunology*, 20: 197–216.

Chernousova, S. and Epple, M. (2013). Silver as antibacterial agent: Ion, nanoparticle, and metal. *Angew Chem. Int. Ed. Engl.*, 52: 1636–53.

Cho, Wan-Seob, Minjung Cho, Jinyoung Jeong, Mina Choi, Hea-Young Cho, Beom Seok Han, Sheen Hee Kim, Hyoung Ook Kim, Yong Taik Lim, Bong Hyun Chung and Jayoung Jeong. (2009). Acute toxicity and pharmacokinetics of 13 nm-sized PEG-coated gold nanoparticles. *Toxicology and Applied Pharmacology*, 236: 16–24.

Das, A., Sinha, M., Datta, S., Abas, M., Chaffee, S., Sen, C.K. and Roy, S. (2015). Monocyte and macrophage plasticity in tissue repair and regeneration. *Am. J. Pathol.*, 185: 2596–606.

Delves, Peter J. and Ivan M. Roitt. (2000). The immune system. *New England Journal of Medicine*, 343: 37–49.

Dewald, Inna, Olga Isakin, Jonas Schubert, Tobias Kraus and Munish Chanana. (2015). Protein identity and environmental parameters determine the final physicochemical properties of protein-coated metal nanoparticles. *The Journal of Physical Chemistry C*, 119: 25482–92.

Díaz, B., Sánchez-Espinel, C., Arruebo, M., Faro, J., de Miguel, E., Magadán, S., Yagüe, C., Fernández-Pacheco, R., Ibarra, M.R., Santamaría, J. and González-Fernández, A. (2008). Assessing methods for blood cell cytotoxic responses to inorganic nanoparticles and nanoparticle aggregates. *Small*, 4: 2025–34.

Dobrovolskaia, M.A. (2015). Pre-clinical immunotoxicity studies of nanotechnology-formulated drugs: Challenges, considerations and strategy. *J. Control Release*, 220: 571–83.

Dobrovolskaia, Marina A. and Scott E. McNeil. (2007). Immunological properties of engineered nanomaterials. *Nat. Nanotechnol.*, 2: 469–78.

Dobrovolskaia, Marina A., Michael Shurin and Anna A. Shvedova. (2016). Current understanding of interactions between nanoparticles and the immune system. *Toxicology and Applied Pharmacology*, 299: 78–89.

Doherty, Gary J. and Harvey T. McMahon. (2009). Mechanisms of endocytosis. *Annu. Rev. Biochem.*, 78: 857–902.

Dunkelberger, Jason R. and Wen-Chao Song. (2010). Complement and its role in innate and adaptive immune responses. *Cell Research*, 20: 34–50.

Eftekhari, Aziz, Solmaz Maleki Dizaj, Leila Chodari, Senem Sunar, Amir Hasanzadeh, Elham Ahmadian and Mohammad Hasanzadeh. (2018). The promising future of nano-antioxidant therapy against environmental pollutants induced-toxicities. *Biomedicine & Pharmacotherapy*, 103: 1018–27.

Erard, Marie, Sophie Dupré-Crochet and Oliver Nüße. (2018). Biosensors for spatiotemporal detection of reactive oxygen species in cells and tissues. *American Journal of Physiology-Regulatory, Integrative and Comparative Physiology*, 314: R667–R83.

Fallarini, Silvia, Tiziana Paoletti, Carolina Orsi Battaglini, Paolo Ronchi, Luigi Lay, Renato Bonomi, Satadru Jha, Fabrizio Mancin, Paolo Scrimin and Grazia Lombardi. (2013). Factors affecting T cell responses induced by fully synthetic glyco-gold-nanoparticles. *Nanoscale*, 5: 390–400.

Fatima, Faria, Preeti Bajpai, Neelam Pathak, Sarika Singh, Shivam Priya and Smita Rastogi Verma. (2015). Antimicrobial and immunomodulatory efficacy of extracellularly synthesized silver and gold nanoparticles by a novel phosphate solubilizing fungus Bipolaris tetramera. *BMC Microbiology*, 15: 52.

Flora, S.J., Mittal, M. and Mehta, A. (2008). Heavy metal induced oxidative stress & its possible reversal by chelation therapy. *Indian J. Med. Res.*, 128: 501–23.

Gagnon, E., Duclos, S., Rondeau, C., Chevet, E., Cameron, P.H., Steele-Mortimer, O., Paiement, J., Bergeron, J.J. and Desjardins. M. (2002). Endoplasmic reticulum-mediated phagocytosis is a mechanism of entry into macrophages. *Cell*, 110: 119–31.

Gahlawat, Geeta, Sristy Shikha, Baldev Singh Chaddha, Saumya Ray Chaudhuri, Shanmugam Mayilraj and Anirban Roy Choudhury. (2016). Microbial glycolipoprotein-capped silver nanoparticles as emerging antibacterial agents against cholera. *Microbial Cell Factories*, 15: 25.

Gasteiger, Georg and Alexander Y. Rudensky. (2014). Interactions between innate and adaptive lymphocytes. *Nature Reviews Immunology*, 14: 631–39.

Gordon, S. (2002). Pattern recognition receptors: doubling up for the innate immune response. *Cell*, 111: 927–30.

Gordon, Siamon. (2016). Phagocytosis: An immunobiologic process. *Immunity*, 44: 463–75.

Gustafson, Heather Herd, Dolly Holt-Casper, David W. Grainger and Hamidreza Ghandehari. (2015). Nanoparticle uptake: The phagocyte problem. *Nano Today*, 10: 487–510.

Hajipour, M.J., Fromm, K.M., Ashkarran, A.A., Jimenez de Aberasturi, D., de Larramendi, I.R., Rojo, T., Serpooshan, V., Parak, W.J. and Mahmoudi, M. (2012). Antibacterial properties of nanoparticles. *Trends Biotechnol.*, 30: 499–511.

Halliwell, B. (2012). Free radicals and antioxidants: Updating a personal view. *Nutr. Rev.*, 70: 257–65.

Halliwell, Barry. (2013). The antioxidant paradox: Less paradoxical now? *Br. J. Clin. Pharmacol.*, 75: 637–44.

Hofman, Kirk, Gautam N. Shenoy, Vincent Chak and Sathy V. Balu-Iyer. (2021). Pharmaceutical aspects and clinical evaluation of COVID-19 vaccines. *Immunological Investigations*, 50: 743–79.

Hrycay, E.G. and Bandiera, S.M. (2015). Involvement of cytochrome P450 in reactive oxygen species formation and cancer. *Adv. Pharmacol.*, 74: 35–84.

Hsueh, Yi-Huang, Kuen-Song Lin, Wan-Ju Ke, Chien-Te Hsieh, Chao-Lung Chiang, Dong-Ying Tzou and Shih-Tung Liu. (2015). The antimicrobial properties of silver nanoparticles in *Bacillus subtilis* are mediated by released Ag+ ions. *PLoS ONE*, 10: e0144306.

Huang, Bill X., Hee-Yong Kim and Chhabil Dass. (2004). Probing three-dimensional structure of bovine serum albumin by chemical cross-linking and mass spectrometry. *Journal of the American Society for Mass Spectrometry*, 15: 1237–47.

Indo, Hiroko P., Hsiu-Chuan Yen, Ikuo Nakanishi, Ken-Ichiro Matsumoto, Masato Tamura, Yumiko Nagano, Hirofumi Matsui, Oleg Gusev, Richard Cornette, Takashi Okuda, Yukiko Minamiyama, Hiroshi Ichikawa, Shigeaki Suenaga, Misato Oki, Tsuyoshi Sato, Toshihiko Ozawa, Daret K. St Clair and Hideyuki J. Majima. (2015). A mitochondrial superoxide theory for oxidative stress diseases and aging. *Journal of Clinical Biochemistry and Nutrition*, 56: 1–7.

Ivask, Angela, Imbi Kurvet, Kaja Kasemets, Irina Blinova, Villem Aruoja, Sandra Suppi, Heiki Vija, Aleksandr Käkinen, Tiina Titma, Margit Heinlaan, Meeri Visnapuu, Dagmar Koller, Vambola Kisand and Anne Kahru. (2014). Size-dependent toxicity of silver nanoparticles to bacteria, yeast, algae, crustaceans and mammalian cells *in vitro*. *PLoS ONE*, 9: e102108.

Iversen, Tore-Geir, Tore Skotland and Kirsten Sandvig. (2011). Endocytosis and intracellular transport of nanoparticles: Present knowledge and need for future studies. *Nano Today*, 6: 176–85.

Iwasaki, Akiko and Ruslan Medzhitov. (2015). Control of adaptive immunity by the innate immune system. *Nature Immunology*, 16: 343–53.

Jamkhande, Prasad Govindrao, Namrata W. Ghule, Abdul Haque Bamer and Mohan G. Kalaskar. (2019). Metal nanoparticles synthesis: An overview on methods of preparation, advantages and disadvantages, and applications. *Journal of Drug Delivery Science and Technology*, 53: 101174.

Janeway, C.A., Jr. (1989). Approaching the asymptote? Evolution and revolution in immunology. *Cold Spring Harb. Symp. Quant. Biol.*, 54 Pt 1: 1–13.

Javed, Rabia, Muhammad Zia, Sania Naz, Samson O. Aisida, Noor ul Ain and Qiang Ao. (2020). Role of capping agents in the application of nanoparticles in biomedicine and environmental remediation: Recent trends and future prospects. *J. Nanobiotechnology*, 18: 172.

Jung, W.K., Koo, H.C., Kim, K.W., Shin, S., Kim, S.H. and Park, Y.H. (2008). Antibacterial activity and mechanism of action of the silver ion in *Staphylococcus aureus* and *Escherichia coli*. *Appl. Environ. Microbiol.*, 74: 2171–8.

Kehrer, J.P. (2000). The Haber-Weiss reaction and mechanisms of toxicity. *Toxicology*, 149: 43–50.

Kumari, Sudha, Swetha Mg and Satyajit Mayor. (2010). Endocytosis unplugged: Multiple ways to enter the cell. *Cell Research*, 20: 256–75.

Lazurko, Caitlin, Erik Jacques, Manuel Ahumada and Emilio I. Alarcon. (2019). Nanomaterials for Its use in biomedicine: An overview. *In*: Emilio I. Alarcon and Manuel Ahumada (eds.). *Nanoengineering Materials for Biomedical Uses* (Springer International Publishing: Cham).

Le Ouay, Benjamin and Francesco Stellacci. (2015). Antibacterial activity of silver nanoparticles: A surface science insight. *Nano Today*, 10: 339–54.

Lech, M. and Anders, H.J. (2013). Macrophages and fibrosis: How resident and infiltrating mononuclear phagocytes orchestrate all phases of tissue injury and repair. *Biochim. Biophys. Acta*, 1832: 989–97.

Lee, Nan-Yao, Wen-Chien Ko and Po-Ren Hsueh. (2019). Nanoparticles in the treatment of infections caused by multidrug-resistant organisms. *Frontiers in Pharmacology*, 10.

Li, Binjing, Ting Zhang and Meng Tang. (2022). Toxicity mechanism of nanomaterials: Focus on endoplasmic reticulum stress. *Science of the Total Environment*, 834: 155417.

Li, J.J., Hartono, D., Ong, C.N., Bay, B.H. and Yung, L.Y. (2010b). Autophagy and oxidative stress associated with gold nanoparticles. *Biomaterials*, 31: 5996–6003.

Li, J.J., Muralikrishnan, S., Ng, C.T., Yung, L.Y. and Bay, B.H. (2010a). Nanoparticle-induced pulmonary toxicity. *Exp. Biol. Med. (Maywood)*, 235: 1025–33.

Liao, Chengzhu, Yuchao Li and Sie Chin Tjong. (2019). Bactericidal and cytotoxic properties of silver nanoparticles. *International Journal of Molecular Sciences*, 20: 449.

Lim, Justin J., Sergio Grinstein and Ziv Roth. (2017). Diversity and versatility of phagocytosis: Roles in innate immunity, tissue remodeling, and homeostasis. *Frontiers in Cellular and Infection Microbiology*, 7.

Ma, D.D. and Yang, W.X. (2016). Engineered nanoparticles induce cell apoptosis: potential for cancer therapy. *Oncotarget*, 7: 40882–903.

Mahla, Ranjeet Singh. (2016). Stem cells applications in regenerative medicine and disease therapeutics. *International Journal of Cell Biology*, 2016: 6940283.

Mailloux, R.J. and Harper, M.E. (2011). Uncoupling proteins and the control of mitochondrial reactive oxygen species production. *Free Radic. Biol. Med.*, 51: 1106–15.

Mailloux, Ryan J. (2015). Teaching the fundamentals of electron transfer reactions in mitochondria and the production and detection of reactive oxygen species. *Redox Biology*, 4: 381–98.

Manke, A., Wang, L. and Rojanasakul, Y. (2013). Mechanisms of nanoparticle-induced oxidative stress and toxicity. *Biomed. Res. Int.*, 2013: 942916.

Martinez, F.O. and Gordon, S. (2014). The M1 and M2 paradigm of macrophage activation: Time for reassessment. *F1000Prime Rep.*, 6: 13.

McMahon, Harvey T. and Emmanuel Boucrot. (2011). Molecular mechanism and physiological functions of clathrin-mediated endocytosis. *Nature Reviews Molecular Cell Biology*, 12: 517–33.

McNamara, Karrina and Syed A.M. Tofail. (2017). Nanoparticles in biomedical applications. *Advances in Physics: X*, 2: 54–88.

Merlot, Angelica M., Danuta S. Kalinowski and Des R. Richardson. (2014). Unraveling the mysteries of serum albumin—more than just a serum protein. *Front. Physiol.*, 5.

Mindell, Joseph A. (2012). Lysosomal acidification mechanisms. *Annual Review of Physiology*, 74: 69–86.

Mitragotri, Samir, Daniel G. Anderson, Xiaoyuan Chen, Edward K. Chow, Dean Ho, Alexander V. Kabanov, Jeffrey M. Karp, Kazunori Kataoka, Chad A. Mirkin, Sarah Hurst Petrosko, Jinjun Shi, Molly M. Stevens, Shouheng Sun, Sweehin Teoh, Subbu S. Venkatraman, Younan Xia, Shutao Wang, Zhen Gu and Chenjie Xu. (2015). Accelerating the translation of nanomaterials in biomedicine. *ACS Nano*, 9: 6644–54.

Mogensen, Trine H. (2009). Pathogen recognition and inflammatory signaling in innate immune defenses. *Clinical Microbiology Reviews*, 22: 240–73.

Napierska, Dorota, Virginie Rabolli, Leen C.J. Thomassen, David Dinsdale, Catherine Princen, Laetitia Gonzalez, Katrien L.C. Poels, Micheline Kirsch-Volders, Dominique Lison, Johan A. Martens and Peter H. Hoet. (2012). Oxidative stress induced by pure and iron-doped amorphous silica nanoparticles in subtoxic conditions. *Chem. Res. Toxicol.*, 25: 828–37.

Nikzamir, Mohammad, Abolfazl Akbarzadeh and Yunes Panahi. (2021). An overview on nanoparticles used in biomedicine and their cytotoxicity. *Journal of Drug Delivery Science and Technology*, 61: 102316.

Nishanth, R.P., Jyotsna, R.G., Schlager, J.J., Hussain, S.M. and Reddanna, P. (2011). Inflammatory responses of RAW 264.7 macrophages upon exposure to nanoparticles: Role of ROS-NFκB signaling pathway. *Nanotoxicology*, 5: 502–16.

Nosaka, Yoshio and Atsuko Y. Nosaka. (2017). Generation and detection of reactive oxygen species in photocatalysis. *Chemical Reviews*, 117: 11302–36.

Nyström, T. (2005). Role of oxidative carbonylation in protein quality control and senescence. *Embo. J.*, 24: 1311–7.

Park, Sung Jean. (2020). Protein-nanoparticle interaction: Corona formation and conformational changes in proteins on nanoparticles. *International Journal of Nanomedicine*, 15: 5783–802.

Pearson, Ryan M., Vanessa V. Juettner and Seungpyo Hong. (2014). Biomolecular corona on nanoparticles: A survey of recent literature and its implications in targeted drug delivery. *Frontiers in Chemistry*, 2.

Petit, A., Mwale, F., Tkaczyk, C., Antoniou, J., Zukor, D.J. and Huk, O.L. (2005). Induction of protein oxidation by cobalt and chromium ions in human U937 macrophages. *Biomaterials*, 26: 4416–22.

Pisoschi, A.M. and Pop, A. (2015). The role of antioxidants in the chemistry of oxidative stress: A review. *Eur. J. Med. Chem.*, 97: 55–74.

Prame Kumar, K., Nicholls, A.J. and Wong, C.H.Y. (2018). Partners in crime: Neutrophils and monocytes/ macrophages in inflammation and disease. *Cell Tissue Res.*, 371: 551–65.

Rai, Mahendra, Avinash P. Ingle, Sonal Birla, Alka Yadav and Carolina Alves Dos Santos. (2016). Strategic role of selected noble metal nanoparticles in medicine. *Critical Reviews in Microbiology*, 42: 696–719.

Redza-Dutordoir, Maureen and Diana A. Averill-Bates. (2016). Activation of apoptosis signalling pathways by reactive oxygen species. *Biochimica et Biophysica Acta (BBA) - Molecular Cell Research*, 1863: 2977–92.

Reichel, Derek, Manisha Tripathi and Manuel Perez, J. (2019). Biological effects of nanoparticles on macrophage polarization in the tumor microenvironment. *Nanotheranostics*, 3: 66–88.

Ren, Guogang, Dawei Hu, Eileen W.C. Cheng, Miguel A. Vargas-Reus, Paul Reip and Robert P. Allaker. (2009). Characterisation of copper oxide nanoparticles for antimicrobial applications. *Int. J. Antimicrob. Agents*, 33: 587–90.

Salem, Wesam, Deborah R. Leitner, Franz G. Zingl, Gebhart Schratter, Ruth Prassl, Walter Goessler, Joachim Reidl and Stefan Schild. (2015). Antibacterial activity of silver and zinc nanoparticles against Vibrio cholerae and enterotoxic Escherichia coli. *International Journal of Medical Microbiology*, 305: 85–95.

Schenten, Dominik and Ruslan Medzhitov. (2011). Chapter 3—The Control of Adaptive Immune Responses by the Innate Immune System. *In*: Frederick W. Alt (ed.). *Advances in Immunology* (Academic Press).

Schumann, J. (2016). It is all about fluidity: Fatty acids and macrophage phagocytosis. *Eur. J. Pharmacol.*, 785: 18–23.

Sharifi, Shahriar, Shahed Behzadi, Sophie Laurent, Laird Forrest, M., Pieter Stroeve and Morteza Mahmoudi. (2012). Toxicity of nanomaterials. *Chemical Society Reviews*, 41: 2323–43.

Shi, Honglian, Laurie G. Hudson and Ke Jian Liu. (2004). Oxidative stress and apoptosis in metal ion-induced carcinogenesis. *Free Radical Biology and Medicine*, 37: 582–93.

Shukla, R.K., Sharma, V., Pandey, A.K., Singh, S., Sultana, S. and Dhawan, A. (2011). ROS-mediated genotoxicity induced by titanium dioxide nanoparticles in human epidermal cells. *Toxicology in vitro: An International Journal Published in Association with BIBRA*, 25: 231–41.

Sies, Helmut, Carsten Berndt and Dean P. Jones. (2017). Oxidative stress. *Annu. Rev. Biochem.*, 86: 715–48.

Sohrabnezhad, Sh, M.J., Mehdipour Moghaddam and T. Salavatiyan. (2014). Synthesis and characterization of CuO–montmorillonite nanocomposite by thermal decomposition method and antibacterial activity of nanocomposite. *Spectrochimica Acta Part A: Molecular and Biomolecular Spectroscopy*, 125: 73–78.

Song, M.F., Li, Y.S., Kasai, H. and Kawai, K. (2012). Metal nanoparticle-induced micronuclei and oxidative DNA damage in mice. *Journal of Clinical Biochemistry and Nutrition*, 50: 211–6.

Steinhardt, Jacinto, Johanna Krijn and Joan G. Leidy. (1971). Differences between bovine and human serum albumins. Binding isotherms, optical rotatory dispersion, viscosity, hydrogen ion titration, and fluorescence effects. *Biochemistry*, 10: 4005–15.

Suzuki, H., Kurihara, Y., Takeya, M., Kamada, N., Kataoka, M., Jishage, K., Ueda, O., Sakaguchi, H., Higashi, T., Suzuki, T., Takashima, Y., Kawabe, Y., Cynshi, O., Wada, Y., Honda, M., Kurihara, H., Aburatani, H., Doi, T., Matsumoto, A., Azuma, S., Noda, T., Toyoda, Y., Itakura, H., Yazaki, Y., Kodama, T. et al. (1997). A role for macrophage scavenger receptors in atherosclerosis and susceptibility to infection. *Nature*, 386: 292–6.

Swanson, J.A. and Hoppe, A.D. (2004). The coordination of signaling during Fc receptor-mediated phagocytosis. *J. Leukoc. Biol.*, 76: 1093–103.

Takeda, Kiyoshi, Tsuneyasu Kaisho and Shizuo Akira. (2003). Toll-like receptors. *Annual Review of Immunology*, 21: 335–76.

Takeuchi, O. and Akira, S. (2010). Pattern recognition receptors and inflammation. *Cell*, 140: 805–20.

Tang, Y., He, R., Zhao, J., Nie, G., Xu, L. and Xing, B. (2016). Oxidative stress-induced toxicity of CuO nanoparticles and related toxicogenomic responses in Arabidopsis thaliana. *Environ. Pollut.*, 212: 605–14.

Taylor, P.R., Martinez-Pomares, L., Stacey, M., Lin, H.H., Brown, G.D. and Gordon, S. (2005). Macrophage receptors and immune recognition. *Annu. Rev. Immunol.*, 23: 901–44.

Thanan, Raynoo, Shinji Oikawa, Yusuke Hiraku, Shiho Ohnishi, Ning Ma, Somchai Pinlaor, Puangrat Yongvanit, Shosuke Kawanishi and Mariko Murata. (2014). Oxidative stress and its significant roles in neurodegenerative diseases and cancer. *International Journal of Molecular Sciences*, 16: 193–217.

Turski, Michelle L. and Dennis J. Thiele. (2009). New roles for copper metabolism in cell proliferation, signaling, and disease. *J. Biol. Chem.*, 284: 717–21.

Valavanidis, Athanasios, Thomais Vlachogianni and Constantinos Fiotakis. (2009). '8-hydroxy-2′-deoxyguanosine (8-OHdG): A critical biomarker of oxidative stress and carcinogenesis. *Journal of Environmental Science and Health, Part C*, 27: 120–39.

Valdiglesias, Vanessa. (2022). Cytotoxicity and genotoxicity of nanomaterials. *Nanomaterials*, 12: 634.

Vance, R.E., Isberg, R.R. and Portnoy, D.A. (2009). Patterns of pathogenesis: Discrimination of pathogenic and nonpathogenic microbes by the innate immune system. *Cell Host Microbe*, 6: 10–21.

Vilanova, Oriol, Judith J. Mittag, Philip M. Kelly, Silvia Milani, Kenneth A. Dawson, Joachim O. Rädler and Giancarlo Franzese. (2016). Understanding the kinetics of protein–nanoparticle corona formation. *ACS Nano*, 10: 10842–50.

Walkey, Carl D., Jonathan B. Olsen, Hongbo Guo, Andrew Emili and Warren C.W. Chan. (2012). Nanoparticle size and surface chemistry determine serum protein adsorption and macrophage uptake. *Journal of the American Chemical Society*, 134: 2139–47.

Wang, Ting-Yi, M., Daben J. Libardo, Alfredo M. Angeles-Boza and Jean-Philippe Pellois. (2017). Membrane oxidation in cell delivery and cell killing applications. *ACS Chemical Biology*, 12: 1170–82.

Widlansky, Michael E. and David D. Gutterman. (2011). Regulation of endothelial function by mitochondrial reactive oxygen species. *Antioxidants & Redox Signaling*, 15: 1517–30.

Xia, Tian, Michael Kovochich, Monty Liong, Jeffrey I. Zink and Andre E. Nel. (2008). Cationic polystyrene nanosphere toxicity depends on cell-specific endocytic and mitochondrial injury pathways. *ACS Nano*, 2: 85–96.

Xie, H., Mason, M.M. and Wise, Sr. J.P. (2011). Genotoxicity of metal nanoparticles. *Rev. Environ. Health*, 26: 251–68.

Yazdi, A.S., Guarda, G., Riteau, N., Drexler, S.K., Tardivel, A., Couillin, I. and Tschopp, J. (2010). Nanoparticles activate the NLR pyrin domain containing 3 (Nlrp3) inflammasome and cause pulmonary inflammation through release of IL-1α and IL-1β. *Proc. Natl. Acad. Sci. USA*, 107: 19449–54.

Zhang, Tong, Matthew J. Gaffrey, Dennis G. Thomas, Thomas J. Weber, Becky M. Hess, Karl K. Weitz, Paul D. Piehowski, Vladislav A. Petyuk, Ronald J. Moore, Wei-Jun Qian and Brian D. Thrall. (2020). A proteome-wide assessment of the oxidative stress paradigm for metal and metal-oxide nanomaterials in human macrophages. *NanoImpact*, 17: 100194.

Zlotoff, Daniel A. and Avinash Bhandoola. (2011). Hematopoietic progenitor migration to the adult thymus. *Annals of the New York Academy of Sciences*, 1217: 122–38.

Chapter 7

Nanomaterials under Microgravity Conditions

Komal Parmar[1] and Jayvadan Patel[2,]*

1. Introduction

Newton discovered the fall of an apple from the tree was due to an attractive force of the earth, i.e., the gravitational force, a fundamental force existing throughout the corners of the known universe. It is defined as an attractive force that acts on the surface of two bodies, that is directly proportional to their masses and inversely proportional to the square of the distance between them. Even though gravity is assumed as present in our world, sometimes it is not required to carry out scientific experiments under its full effect. In these cases, experiments are conducted under microgravity conditions in which the influence of gravity is greatly reduced. We are all familiar with weightless astronauts floating around during space exploration, but there are numerous repercussions of microgravity that are yet unknown. Microgravity has an impact on the behavior of materials, the progression of chemical reactions, and the functioning of organisms. Microgravity can also provide a unique environment for grounded scientists to better comprehend the fundamental features of matter, like the interaction of microscopic and nanoscopic particles. Material science experiments have been the most discussed topics. The latest developments in aerospace science have permitted to study the microgravity in laboratories on Earth (Herranz et al. 2013, Aceto et al. 2016, Li et al. 2017, Anken et al. 2017, Kiss et al. 2019). The first experiments under zero gravity were reported during the fly-back phase of the Apollo 14 moon mission in 1971 (Sharpe and Wright 2009).

Microgravity, also known as micro-g environment, refers to the condition of zero gravity, but this is untrue (Figure 1). In reality, gravity forces can never be precisely

---

[1] ROFEL, Shri G.M. Bilakhia College of Pharmacy, Vapi-396191, Gujarat, India.
[2] Aavis Pharmaceuticals, Hoschton, USA, and Sankalchand Patel University, Gujarat, India.
* Corresponding author: jayvadan04@yahoo.com

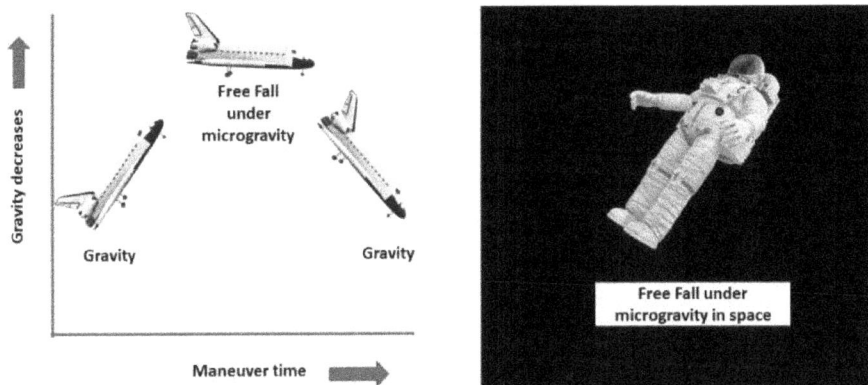

Figure 1. Schematic diagram to understand microgravity.

zero but are just minimal. Thus, people or objects experience weightlessness in space under the influence of microgravity. In this condition, astronauts can have spacewalk and float inside a spaceship. All the objects carried by spaceships are in the state of free fall along with the spacecraft and thus experience weightlessness. Likewise, cosmonauts can easily move hefty objects in space under the influence of microgravity, as the experienced environmental conditions are different than the Earth's. Consequently, it turns into stress on humans and objects.

Studies of microgravity science have created new opportunities for physical and life science research that are important to develop a new understanding of how microgravity works, responds, and functions in physical structures and the human body to be properly equipped for a trip to human-crewed missions in space. Various experiments were conducted using simulators to mimic the space environment and study the impact of microgravity. Few studies have revealed that micro-g modifies the buildup of organic (Robet et al. 2019, Devara et al. 2015) and inorganic materials (Merzhanov et al. 2001). In organic change, microgravity may influence biological processes such as cell proliferation, differentiation, metabolism, and other functions of the cells (Bizzarri 2014, Gioia et al. 2018, Mouhamad et al. 2019, Fukazawa et al. 2019, Grimm 2019, Yatagai et al. 2019, Grimmet et al. 2020, Shi et al. 2020).

A microgravity atmosphere provides an unprecedented facility in which experts can study and investigate physical events, phenomena, and processes that are typically obscured by the Earth's gravitational force. For instance, crystal growth (Sorgenfrei 2018, Hayakawa et al. 2017), capillary-driven flow (Li et al. 2015), and diffusive transport process (Yan et al. 2007) can be influenced by normal gravitational forces, which can be easily studied under microgravity conditions. Microgravity provides accurate measurements of material properties such as viscosity and surface tension (Mohr et al. 2019). Apart from various applications of microgravity research on materials (Table 1), the zero-gravity environment also influences various properties of materials. Few of them are discussed in brief in the following review.

Table 1. Applications of microgravity-based research.

Application of Microgravity	Reference
3D-printing of metal components	Zocca et al. 2019
Container-less processing of metals and alloys	Rathz et al. 2001
High-quality crystal growth	Benz and Dold 2002, Corregidor et al. 2002
Bulk crystal growth	Arivanandhan et al. 2012
Thermophysical property accurate measurement	Egry 2001, Ishikawa et al. 2015
Fluid processing	Wuest et al. 2017
Protein crystal growth	McPherson and DeLucas 2015
Biomedical applications	Mitsuhara et al. 2013, Costantini et al. 2019

2. Influence of Microgravity on various Parameters

Microgravity impacts material processing in several manners, involving changes in basic characteristics of either material or processes. Furthermore, we will see a brief detail of the impact of microgravity on various factors and phenomena.

2.1 Growth Parameter

Crystal growth is the part of the crystallization process wherein the addition of atoms, ions, or polymer strings get attached to the crystalline lattice. Through a variety of applications, space-based crystal formation research has the potential to significantly improve life on Earth. Microgravity crystals are typically larger and more solid or organized than those formed on Earth (McPherson and DeLucas 2015). For instance, a higher growth rate of a ternary alloy (Indium-Gallium-Antimony) was observed under the influence of microgravity. Briefly, the growth and dissolution were affected mainly by steady-state equilibrium in the melt composition because of the convection under normal gravity. Microgravity suppresses the convection and influences the complex heat and mass transfer in the melt. This non-steady-state of equilibrium in melt composition under microgravity aided the enhanced crystal growth of the ternary alloy (Yu et al. 2019). Recently, an investigation was carried out to study the impact of the microgravity environment on the crystal growth of ettringite $[Ca_3Al(OH)_6 \cdot 12H_2O]2 \cdot (SO_4)_3 \cdot 2H_2O$. Briefly, $Ca(OH)_2/Al_2(SO_4)_3$ solutions were combined and reacted for 10 seconds, followed by quick filtration of the suspension and successive quenching with acetone. Results demonstrated precipitation of smaller crystals (1–2.9 μm) with slightly different morphology in larger amounts than those in normal gravity (1–3.5 μm). The aspect ratio was found to be similar to that of crystals grown under normal gravity. The smaller size of ettringite crystals might be associated with the altered transport of ions to the crystal surface. The absence of convection leads to the generation of more nuclei as ion transport is facilitated by the diffusion mechanism (Meier et al. 2020). An experimental setup was carried out at the International Space Station (ISS) to study the effect of microgravity on microstructural cementation of tri-calcium silicate (Ca_3Si) paste. The results demonstrated that microgravity modifies many features of the microstructure of high water-to-cement ratio Ca₃Si paste. The admixture was found to be more porous, attributed to the lack of buoyancy in the microgravity environment. Elongated-plate-like crystallites were found in the microgravity environment, which might be due to additional autonomy to grow into a microstructure driven by chemical diffusion, deprived of any complications from the gravitational effects. Microgravity promotes diffusion-controlled growth and even solidification of microstructures (Neves et al. 2019). In an experimental setup, InxGa1−xSb crystals were grown at the ISS under microgravity conditions, and a similar protocol was carried out under earth gravity conditions using the vertical gradient freezing method. The shape of the growth interface of the crystals was nearly flat under microgravity, whereas under normal gravity, it was exceedingly concave, with the initial seed interface being almost even and having facets at the edges. The quality of the microgravity crystals was found to be better than that of the normal gravity samples. Also, the growth rate was higher under microgravity when compared with normal gravity (Inatomi et al. 2015).

2.2 Bubble Formation

Bubble phenomena are observed in a process such as boiling, cavitation, crystal growth, and chemical reactions involving gas and liquid in stirring vessels. In microgravity conditions, bubble formation is influenced by other forces such as pressure, viscosity, and surface tension (Ronshin et al. 2021, Liang et al. 2011, Obreschkow et al. 2006). Lin and co-workers essayed numerical simulations of bubble dynamics under the influence of microgravity. The study briefly demonstrates the coalescence process of two bubbles under microgravity conditions. Results revealed different conditions for two bubbles under different gravity conditions. For instance, it was observed that the velocity fields inside and around the bubbles under different gravity conditions were almost the same. In contrast, the strength of vortices behind the bubbles was weaker under microgravity conditions. Furthermore, it was also found that under microgravity conditions, two bubbles took a long time for coalescence with reduced deformations (Lin et al. 2008). In other investigations, it was concluded that foams are quite stable under the stimulus of microgravity. It was supported by the fact that, under normal gravity conditions, foam is subjected to free drainage wherein the liquid films become thin and eventually break. This phenomenon is absent in a microgravity environment, and thus foams containing a large number of liquids can be studied for longer periods (Langevin 2017, Caps et al. 2014, Langevin and Vignes-Adler 2014, Vandewalle et al. 2011). In one work, researchers demonstrated the investigation of foams containing solid particles both in normal and micro-gravity conditions. Foam containing silicon dioxide nanoparticles were subjected to varied gravity conditions. Results illustrated similar deformation of foam bubbles under any type of gravity level. However, it was observed that the volume of solid-containing foam was increased under microgravity levels (Somosvari et al. 2011).

2.3 Convective Heat Transfer

This quantity is defined as the transfer of heat from one place to another with the movement of the fluid and is affected by several aspects, including various fluid properties, the presence of a thermal gradient, geometric configuration, flow condition, and gravity. The effect of gravity on the boiling heat transfer efficiency becomes a critical concern in aerospace applications since the temperature control systems are under fluctuating gravity conditions (Iceri et al. 2020). A pool boiling experiment was conducted onboard flights to study the effect of gravity on heat transfer. Ethanol was used at low pressures of 0.014–0.017 MPa. Initially, enhancement in heat transfer is observed in microgravity which might be due to the extension of a microlayer underneath enlarged bubbles for liquid, while later, a reduction in heat transfer was reported due to the growth of primary bubbles at the base of a large and coalesced bubble. Thus, the impact of gravity on heat transfer was not decisive since there was an opposing trend in investigational results (Ohta and Baba 2013). Warrier and others carried out experiments on pool boiling on an aluminum wafer with five artificial nucleation cavities in the boiling experimental facility onboard the ISS. Perfluoro-n-hexane was used as the test liquid. After commencement of boiling, the bubbles generated coalesced into a large bubble in the middle of the heater. As the heat flux increased, the large bubble shifted from the heater, but floated on the surface, and small bubbles nucleated on the wall surface continued to amalgamate with the large bubble. The heat flux measured at steady state nucleate boiling was found to be lower than in normal gravity conditions (Warrier et al. 2015). In another such experiment, the impact of reduced gravity on convective heat transfer was studied by Lotto and co-researchers. A series of tests were performed to assess convection under reduced-gravity conditions averaging 0.45 m·s^{-2} (0.05 g) attained onboard a parabolic aircraft. The results exhibited a decline in net heat transfer of around 61% in-flight compared to a 1 g terrestrial baseline by means of the same experimental arrangement. The mean experimental Nusselt Number of 19.05 was obtained, which statistically correlated with the projected value of 18.90, estimated using the Churchill-Chu correlation for free convective heat transfer from a finite, flat, vertical plate. Inferring this to analogous performance in true microgravity (10^{-6} g) shows that these conditions should yield a Nusselt Number of 1.27, which is 2.6% of the scale of free convection at 1 g, or depicts a reduction of 97.4% (Lotto et al. 2017).

2.4 Solidification

Solidification is the most important step in the fabrication process involving liquid and solids. At the dynamic solid-liquid interface, some complex physical processes take place. The solidification of melt material is often influenced by gravity, which affects the formation of the final product. On Earth, gravity uses natural convection process to solidify the melt material. In addition, density differences will lead to the separation of solid from liquid. From the studies, it is observed that microgravity removes these effects and ultimately provides a more qualitative solidification process (Nguyen-Thi et al. 2017). Recently, an experiment was carried out on board the Tiangong 2 space laboratory of China under a microgravity environment. A solidification experiment was carried out with Aluminium-Bismuth-Tin immiscible alloy. The results demonstrated the absence of any cavity in the final microstructure composition, which might be due to the diminished activity of convective flow of the melt and the Stokes motion of the droplets and gas bubbles under microgravity. This, in turn, was supportive in suppressing the occurrence of macro-segregation, thereby preventing the porosity formation. Further, a microgravity environment prevented the bonding of melt and wall of the crucible, thereby promoting nucleation (Jiang et al. 2019). In another setup, the impact of gravity on the solidification microstructure of Al–Al$_3$Ni eutectic alloy was studied by means of a 50-m-long drop tube. It was observed that the average inter-rod spacing was always higher at different growth rates under microgravity than compared to normal gravity. Furthermore, the spacing difference between normal gravity and microgravity samples diminished gradually with increasing growth rates (Zhang et al. 2014).

2.5 Buoyancy

This phenomenon is defined as the tendency to float on fluid, whereas sedimentation is to settle down at the bottom. On earth, heavy things will sediment at the bottom, and light things will float on the fluid due to the relative difference in densities, wherein microgravity promotes the uniform distribution of particles by minimizing sedimentation and buoyancy effect. In an experiment, tin particles were studied for their distribution in the liquid phase. A sample of 10% w/v of tin phase was held for 10 h and 10 min at 185°C in the liquid phase, on the ground means on Earth under gravitational force and another at the Microgravity Glovebox at the ISS. It was observed that tin particles remained evenly distributed under microgravity conditions, while under earth gravity, tin particles remained sedimented on the top of the sample as tin particles were less dense than the liquid. The absence of sedimentation leads to homogeneous tin particle distribution in the liquid phase. Figure 2 demonstrates a schematic representation of the result for easy interpretation (Gulsoy et al. 2011).

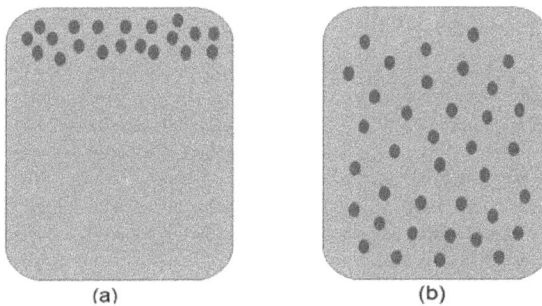

(a)   (b)

Figure 2. Schematic representation of tin particle distribution under the influence of normal gravity (a) and microgravity (b).

Outlook

Unlike other stress conditions, microgravity itself poses a stressful environment for the materials used in space research. Several physical parameters, but not limited to crystal formation, shape, size of crystals, convection, solidification of melt material, and bubble dynamics are affected by exposure to microgravity. The study of microgravity from a nanoscale perspective is encouraged because few investigations have been documented. The basis for developing novel nanomaterials for space-based construction, repair, and manufacture is an understanding of formation from the bottom up under the influence of microgravity. Understanding the impact of microgravity on many physicochemical properties of materials can help to use the microgravity environment for the beneficial purpose of involving such materials. Experiments in a microgravity environment have the potential to yield findings that will benefit living beings on Earth while also furthering our understanding of space.

References

Aceto, J., Nourizadeh-Lillabadi, R., Bradamante, S., Maier, J.A., Alestrom, P., van Loon, J.J.W.A. and Muller, M. (2016). Effects of microgravity simulation on zebrafish transcriptomes and bone physiology—exposure starting at 5 days post fertilization. *NPJ Microgravity*, 2: 16010. https://doi.org/10.1038/npjmgrav.2016.10.

Anken, R., Knieand, M. and Hilbig, R. (2017). Inner ear otolith asymmetry in late-larval cichlid fish (Oreochromis mossambicus, Perciformes) showing kinetotic behaviour under diminished gravity. *Sci. Rep.*, 7: 15630. https://doi.org/10.1038/s41598-017-15927-z.

Arivanandhan, M., Rajesh, G., Tanaka, A., Ozawa, T., Okano, Y., Inatomi, Y. and Hayakawa, Y. (2012). Bulk growth of InGaSb alloy semiconductor under terrestrial conditions: A preliminary study for microgravity experiments at ISS. *Defect and Diffusion Forum*, 323-325: 539–544. https://doi.org/10.4028/www.scientific.net/ddf.323-325.539.

Benz, K.W. and Dold, P. (2002). Crystal growth under microgravity: Present results and future prospects towards the International Space Station. *J. Cryst. Growth*, 237-239(3): 1638–1645.

Bizzarri, M. (2014). Systems biology and microgravity effects on living organisms. *Curr. Synthetic Sys. Biol.*, 2: 3. 10.4172/2332-0737.1000e111.

Caps, H., Vandewalle, N., Saint-Jalmes, A., Saulnier, L., Yazhgur, P., Rio, E., Salonen, A. and Langevin, D. (2014). How foams unstable on Earth behave in microgravity? *Colloids Surf. A Physicochem. Eng. Asp.*, 457: 392–396.

Corregidor, V., Babentsov, V., Castano, J.L., Fiederle, M., Feltgen, T., Benz, K. and Dieguez, E. (2002). Characterization of CdTe:Zn:V crystals grown under microgravity conditions. *J. Mater. Res.*, 17(12): 3037–3041.

Costantini, D., Overi, D., Casadei, L., Cardinale, V., Nevi, L., Carpino, G., Di Matteo, S., Safarikia, S., Valerio, M., Melandro, F., Bizzarri, M., Manetti, C., Berloco, P.B., Gaudio, E. and Alvaro, D. (2019). Simulated microgravity promotes the formation of tridimensional cultures and stimulates pluripotency and a glycolytic metabolism in human hepatic and biliary tree stem/progenitor cells. *Sci. Rep.*, 9: 5559. https://doi.org/10.1038/s41598-019-41908-5.

Devarayan, K., Yesupatham, S., Yang, S.L. and Byoung-Suhk, K. (2015). Effect of microgravity on fungistatic activity of an α-aminophosphonate chitosan derivative against *Aspergillus niger*. *PLoS ONE*, 10(10): e0139303. doi:10.1371/journal.pone.0139303.

Egry, I. (2001). Thermophysical property measurements of liquid metals in microgravity. *AIP Conference Proceedings*, 552: 325. https://doi.org/10.1063/1.1357942.

Fukazawa, T., Tanimoto, K., Shrestha, L., Imura, T., Takahashi, S., Sueda, T., Hirohashi, N., Hiyama, E. and Yuge, L. (2019). Simulated microgravity enhances CDDP-induced apoptosis signal via p53-

independent mechanisms in cancer cells. *PLoS ONE*, 14(7): e0219363. doi.org/10.1371/journal. pone.0219363.

Gioia, M., Michaletti, A., Scimeca, M., Marini, M., Tarantino, U., Zolla, L. and Coletta, M. (2018). Simulated microgravity induces a cellular regression of the mature phenotype in human primary osteoblasts. *Cell Death Discov.*, 4(59). doi.org/10.1038/s41420-018-0055-4.

Grimm, D. (2019). Guest edited collection: Gravitational biology and space medicine. *Sci. Rep.*, 9: 14399. doi.org/10.1038/s41598-019-51231-8.

Grimm, D., Wehland, M., Corydon, T.J., Richter, P., Prasad, B., Bauer, J., Egli, M., Kopp, S., Lebert, M. and Kruger, M. (2020). The effects of microgravity on differentiation and cell growth in stem cells and cancer stem cells. *Stem Cells Transl. Med.*, 9(8): 882–894.

Gulsoy, E.B., Wittman, K., Thompson, J. and Voorhees, P.W. (2011). Coarsening in solid-liquid mixtures: Effect of microgravity accelerations on particle sedimentation. *49th AIAA Aerospace Sciences Meeting including the New Horizons Forum and Aerospace Exposition, Orlando, Florida.* https://doi.org/10.2514/6.2011-1346.

Herranz, R., Anken, R., Boonstra, J., Braun, M., Christianen, P.C., de Geest, M., Hauslage, J., Hilbig, R., Hill, R.J., Lebert, M., Medina, F.J., Vagt, N., Ullrich, O., van Loon, J.J. and Hemmersbach, R. (2013). Ground-based facilities for simulation of microgravity: organism-specific recommendations for their use, and recommended terminology. *Astrobiology*, 13(1): 1–17.

Iceri, D.M., Zummo, G., Saraceno, L. and Ribatski, G. (2020). Convective boiling heat transfer under microgravity and hypergravity conditions. *Int. J. Heat Mass Transfer*, 153: 119164.

Inatomi, Y., Sakata, K., Arivanandhan, M., Rajesh, G., Nirmal Kumar, V., Koyama, T., Momose, Y., Ozawa, T., Okano, Y. and Hayakawa, Y. (2015). Growth of InxGa1–xSb alloy semiconductor at the International Space Station (ISS) and comparison with terrestrial experiments. *NPJ Microgravity*, 1: 15011. https://doi.org/10.1038/npjmgrav.2015.11.

Ishikawa, T., Okada, J., Paradis, P.F., Watanebe, Y. and Watanebe, M. (2015). Surface tension and viscosity measurement of highly viscous melts using a sample rotation. *International Journal of Microgravity Science and Application*, 32(1): 320106. https://doi.org/10.15011/jasma.32.1.320106.

Jiang, H., Li, S., Zhang, L., He, J. and Zhao, J. (2019). Effect of microgravity on the solidification of aluminum–bismuth–tin immiscible alloys. *NPJ Microgravity*, 5(1): 26. https://doi.org/10.1038/s41526-019-0086-z.

Kiss, J.Z., Wolverton, C., Wyatt, S.E., Hasenstein, K.H. and van Loon, J.J.W.A. (2019). Comparison of microgravity analogs to spaceflight in studies of plant growth and development. *Front Plant Sci.*, 10: 1577. https://doi.org/10.3389/fpls.2019.01577.

Langevin, D. (2017). Aqueous foams and foam films stabilised by surfactants. Gravity-free studies. *Comptes Rendus Mécanique*, 345(1): 47–55.

Langevin, D. and Vignes-Adler, M. (2014). Microgravity studies of aqueous wet foams. *Eur. Phys. J. E.*, 37: 16. doi:10.1140/epje/i2014-14016-3.

Li, X., Anken, R., Liu, L., Wang, G. and Liu, Y. (2017). Effects of simulated microgravity on otolith growth of larval zebrafish using a rotating-wall vessel: Appropriate rotation speed and fish developmental stage. *Microgravity Sci. Technol.*, 29: 1–8.

Li, Y., Hu, M., Liu, L., Su, Y.Y., Duan, L. and Kang, Q. (2015). Study of capillary driven flow in an interior corner of rounded wall under microgravity. *Microgravity Sci. Tec.*, 27: 193–205.

Liang, R., Liang, D., Yan, F., Liao, Z. and Duan, G. (2011). Bubble motion near a wall under microgravity: Existence of attractive and repulsive forces. *Microgravity Sci. Technol.*, 23: 79–88. doi.org/10.1007/s12217-010-9238-1.

Lin, P., Huixiong, L., Xiao, L., Jianfu, Z., Tingkuan, C. and Yuqin, Z. (2008). Numerical simulation of bubble dynamics in microgravity. *Microgravity Sci. Technol.*, 20(3-4): 247–251.

Lotto, M.A., Johnson, K.M., Nie, S.W. and Klaus, D.M. (2017). The impact of reduced gravity on free convective heat transfer from a finite, flat, vertical plate. *Microgravity Sci. Tec.*, 29: 371–379.

McPherson, A. and DeLucas, L.J. (2015). Microgravity protein crystallization. *NPJ Microgravity*, 1: 15010. doi.org/10.1038/npjmgrav.2015.10.

Meier, M.R., Lei, L., Rinkenburger, A. and Plank, J. (2020). Crystal growth of [Ca3Al(OH)6·12H2O]2·(SO4)3·2H2O (Ettringite) studied under microgravity conditions. *J. Wuhan. Univ. Technol.-Mat. Sci. Edit.*, 35: 893–899.

Merzhanov, A.G., Rogachev, A.S., Rumanov, E.N., Sanin, V.N., Sytchev, A.E., Shcherbakov, V.A. and Yukhvid, V.I. (2001). Influence of microgravity on self-propagating high-temperature synthesis of refractory inorganic compounds. *Cosm. Res.*, 39: 210–223.

Mitsuhara, T., Takeda, M., Yamaguchi, S., Manabe, T., Matsumoto, M., Kawahara, Y., Yuge, L. and Kurisu, K. (2013). Simulated microgravity facilitates cell migration and neuroprotection after bone marrow stromal cell transplantation in spinal cord injury. *Stem Cell Res. Ther.*, 4: 35. https://doi.org/10.1186/scrt184.

Mohr, M., Wunderlich, K.R., Zweiacker, K., Prades-Rödel, S., Sauget, R., Blatter, A., Logé, R., Dommann, A., Neels, A., Johnson, W.L. and Fecht, H.J. (2019). Surface tension and viscosity of liquid $Pd_{43}Cu_{27}Ni_{10}P_{20}$ measured in a levitation device under microgravity. *NPJ Microgravity*, 5: 4. doi: 10.1038/s41526-019-0065-4.

Mouhamad, R.S., Shallal, H.H. and Al-Daoude, A. (2019). Microgravity effects on the growth, cell cytology properties and DNA alterations of three iraqi local plants. *J. Rice Res.*, 7(2): 207. 10.4172/2375-4338.1000207.

Neves, J.M., Collins, P.J., Wilkerson, R.P., Grugel, R.N. and Radlinska, A. (2019). Microgravity effect on microstructural development of Tri-calcium Silicate (C3S) paste. Front. Mater., 6: 83. https://doi.org/10.3389/fmats.2019.00083.

Nguyen-Thi, H., Reinhart, G. and Billia, B. (2017). On the interest of microgravity experimentation for studying convective effects during the directional solidification of metal alloys. *Comptes Rendus Mécanique*, 345(1): 66–77.

Obreschkow, D., Kobel, P., Dorsaz, N., de Bosset, A., Nicollier, C. and Farhat, M. (2006). Cavitation bubble dynamics inside liquid drops in microgravity. *Phys. Rev. Lett.*, 97(9): 094502. doi: 10.1103/PhysRevLett.97.094502.

Ohta, H. and Baba, S. (2013). Boiling experiments under microgravity conditions. Exp. Heat Transf., 26(2-3): 266–295.

Rathz, T.J., Robinson, M.B., Li, D., Workman, G.L. and Williams, G. (2001). Study of the containerless undercooling of Ti-Ce immiscible alloys. *Journal of Materials Science*, 36: 1183–1188. https://doi.org/10.1023/A:1004885926618.

Roberts, D.R., Brown, T.R., Nietert, P.J., Eckert, M.A., Inglesby, D.C., Bloomberg, J.J., George, M.S. and Asemani, D. (2019). Prolonged microgravity affects human brain structure and function. *Am. J. Neuroradiol.* doi: 10.3174/ajnr.A6249.

Ronshin, F., Sielaff, A., Tadrist, L., Stephan, P. and Kabov, O. (2021). Dyanmics of bubble growth during boiling at microgravity. *J. Phys.: Conf. Ser.*, 2119 012170. doi:10.1088/1742-6596/2119/1/012170.

Sharpe, R.J. and Wright, M.D. (2009). Analysis of microgravity experiments conducted on the appollo spacecraft. *NASA*/TM-2009-215905.

Shi, L., Tian, H., Wang, P., Li, L., Zhang, Z., Zhang, J. and Zhao, Y. (2020). Spaceflight and simulated microgravity suppresses macrophage development via altered RAS/ERK/NFκB and metabolic pathways. *Cell Mol. Immunol.* https://doi.org/10.1038/s41423-019-0346-6.

Somosvári, B.M., Bárczy, P., Szóke, J., Szirovicza, P. and Bárczy, T. (2011). Focus: Foam evolution and stability in microgravity. *Colloids Surf A Physicochem. Eng. Asp.*, 382: 58–63.

Sorgenfrei, T. (2018). 30 years of crystal growth under microgravity conditions in freiburg: an overview of past activities. *Cryst. Res. Technol.*, 53(5): 1700265. doi.org/10.1002/crat.201700265.

Vandewalle, N., Caps, H., Delon, G., Saint-Jalmes, A., Rio, E., Saulnier, L., Adler, M., Biance, A.L., Pitois, O., Cohen Addad, S., Hohler, R., Weaire, D., Hutzler, S. and Langevin, D. (2011). Foam stability in microgravity. *J. Phys. Conf. Ser.*, 327: 012024. doi:10.1088/1742-6596/327/1/012024.

Warrier, G.R., Dhir, V.K. and Chao, D.F. (2015). Nucleate Pool Boiling eXperiment (NPBX) in microgravity: International space station. *Int. J. Heat Mass Transf.*, 83: 781–798.

Wuest, S.L., Stern, P., Casartelli, E. and Egli, M. (2017). Fluid dynamics appearing during simulated microgravity using random positioning machines. *PLoS One*, 12(1): e0170826. doi: 10.1371/journal.pone.0170826.

Yan, Y., Pan, S., Jules, K. and Saghir, M.Z. (2007). Vibrational effect on thermal diffusion under different microgravity environments. *Microgravity Sci. Tec.*, 19: Article 12. https://doi.org/10.1007/BF02911863.

Yasuhuro, H., Velu, N.K., Mukannan, A., Govindasamy, R., Tadanobu, K., Yoshimi, M., Kaoruho, S., Tetsuo, O., Yasunori, O. and Yuko, I. (2017). Effects of gravity and crystal orientation on the growth of InGaSb ternary alloy semiconductors. *International Journal of Microgravity Science and Application*, 34(1): 340111. doi.org/10.15011//jasma.34.340111.

Yatagai, F., Honma, M., Dohmae, N. and Ishioka, N. (2019). Biological effects of space environmental factors: A possible interaction between space radiation and microgravity. *Life Sci Space Res.*, 20: 113–123.

Yu, J., Inatomi, Y., Kumar, V.N., Hayakawa, Y., Okano, Y., Arivanandhan, M., Momose, Y., Pan, X., Liu, Y., Zhang, X. and Luo, X. (2019). Homogeneous InGaSb crystal grown under microgravity using Chinese recovery satellite SJ-10. *NPJ Microgravity*, 5: 8. doi.org/10.1038/s41526-019-0068-1.

Zhang, N., Xinghong, L., Shaobo, F. and Yuhu, R. (2014). Mechanism of gravity effect on solidification microstructure of eutectic alloy. *J. Mater. Sci. Technol.*, 30(5): 499–503.

Zocca, A., Lüchtenborg, J., Mühler, T., Wilbig, J., Mohr, G., Villatte, T., Léonard, F., Nolze, G., Sparenberg, M., Melcher, J., Hilgenberg, K. and Günster, J. (2019). Enabling the 3D printing of metal components in μ-gravity. *Adv. Mater. Technol.*, 4(10): 1900506. https://doi.org/10.1002/admt.201900506.

Chapter 8

Nanomaterials under High-Pressure Conditions

Denis Machon[1,2]

1. Introduction

Nanosized materials show extensive new physical and chemical properties compared to bulk samples. It is now well established that nanoscience is the base for a new technological revolution, and, actually, some nanomaterials are already used for industrial applications (Chen and Mao 2007). Therefore, they have been the focus of extensive research interest in the attempt to understand, control, and tune the stability and the properties of functional nanomaterials.

At the nanoscale, the contribution of the surface energy in the energetic balance plays a decisive role. This additional energy may modify the phase equilibrium and stabilize new structures at ambient conditions with properties different from the bulk counterparts (Roduner 2006). For instance, bulk TiO_2 adopts a rutile structure, whereas TiO_2 nanoparticles crystallize in an anatase structure (Navrotsky 2003). These nanoparticles are widely used for their photocatalytic properties (Diebold 2003).

To better understand the effect of surface energy on phase stability, the combination of pressure and particle size is particularly important as, keeping the particle size constant, the pressure allows the energy landscapes of the system to be explored. Note that there are some reported cases of pressure-induced coalescence for which this approach is inadequate as pressure application induces a drastic change in the geometry and the size of the nanostructures (Bai et al. 2019). However, such

---

[1] Institut Lumière Matière, CNRS, UMR 5306, Université Lyon 1, Campus LyonTech - La Doua, Bâtiment Brillouin, 10 rue Ada Byron, 69622 Villeurbanne CEDEX, France.

[2] Laboratoire Nanotechnologies Nanosystèmes (LN2) - CNRS IRL-3463 Institut Interdisciplinaire d'Innovation Technologique (3IT), Université de Sherbrooke, 3000 Boulevard Université, Sherbrooke, J1K OA5 Québec, Canada.

Email: denis.machon@univ-lyon1.fr

examples are scarce. In general, no cold sintering occurs with nanoparticles. This can be demonstrated by monitoring the confined acoustic vibrations of the nanoparticles under high stresses using *in situ* low-frequency Raman spectroscopy (Saviot et al. 2012, Girao et al. 2017).

In addition, pressure and particle size are two parameters that can be used conjointly to stabilize new phases (Roduner 2006, Tolbert and Alivisatos 1995, San Miguel 2006, Navrotsky 2003). Nanoscale can also bring topological changes, which can induce drastic modifications in physical properties. As such, nanotube-shaped objects, nanospheres, and bidimensional sheets of the same bonded atoms can exhibit totally different behaviors.

High-pressure science was developed in the 50s with the advent of new devices to generate pressure above the GPa range (Jayaraman 1986). For example, Diamond-Anvil Cells (DAC) allow reaching tens of GPa, keeping optical access and allowing to perform spectroscopic measurements. By now, pressures from 1 to 400 GPa (Shen and Mao 2017) or more can be generated, and with the advent of synchrotron radiation facilities, structural changes can be monitored under such extreme pressures (Shen and Mao 2017).

The sample, immersed in a compressible fluid known as the pressure transmitting medium (PTM), is placed in a compression chamber, formed by a pierced metal gasket, and squeezed between the polished culets of two opposed diamonds. The fluid PTM is chosen according to the desired temperature and pressure, and a potential physical or chemical interaction with the sample. The dominance of surface atoms in nanosystems and nanomaterials makes the interaction with the PTM a major issue in high-pressure experiments.

These technological developments paved the way for high-pressure investigations of materials with two main advantages:

1) To study the matter as a function of pressure (and temperature). In that case, pressure is a major thermodynamic parameter that can induce phase transformation or chemical reactions.

2) The response of materials to pressure allows exploring their mechanical properties. The pressure range allows for determining the bulk modulus of materials and their behavior under very high stress. In the case of nanostructures, it also permits to tune the topology of the materials.

In this chapter, we will discuss both aspects. First, the interest in pressure-induced transition in nanomaterials such as oxide nanoparticles will be developed. The shift of the transition line, the appearance of metastable states, etc., can be discussed using increasingly sophisticated thermodynamic models. These latter include interface energy but also the effect of the point defects inherent to the nanostructure. The second part will be dedicated to the mechanical properties of graphene, the archetype of 2D materials. The mechanism of stress transmission to an atomically-thin material will be discussed, and we will present a practical example that aims to determine the coupling between components in graphene-based nanocomposites.

2. Pressure-induced Phase Transitions in Nanomaterials

Phase stability in finite-sized materials can be modified by surface energy control. For instance, under ambient conditions, in nanostructures, polymorphs that differ from the bulk counterpart may be stabilized. This is typically a size-induced phase transition as reported for nickel (Illy et al. 1999), cobalt (Ram 2001), or various oxides such as ZrO_2 or TiO_2 (Navrotsky 2001, Navrotsky 2003).

However, if the concept of surface energy exists to describe free nanoparticle properties, this model can hardly be generalized to describe phase transitions, in particular, in high-pressure experiments. As a matter of fact, nanoparticles are embedded in a medium that creates interface energy. In general, the nature of this interface (chemical interaction, defects, etc.) contributes to and potentially modifies the phase equilibria. Exploring the pressure-induced phase transitions in nanomaterials is a way to estimate the contribution of interface energy to phase stability at the nanoscale.

Therefore, the interest in studying nanoparticles under high pressure is at least two-fold: (i) to gain a fundamental understanding of the thermodynamics when the interfacial energy becomes of the same magnitude as the intraparticle energy, and (ii) to stabilize new structures that may have a potential interest as functional materials.

When dealing with nanocrystals under pressure, two main behaviors are generally reported:

(1) If a pressure-induced phase transition occurs between phase I and phase II in a bulk sample, the transition pressure changes (generally increases) when decreasing the particle size.

(2) New phases and states (new polymorphs or amorphous states) compared to the bulk can be observed in the (size, pressure) phase diagram.

One of the first examples was reported in CdSe, for which a $1/R$ variation of the transition pressure was observed, where R is the particle size (Tolbert and Alivisatos 1995). This effect can be described by considering the appropriate thermodynamics potential, including the interface energy contribution and the tendency of the high-pressure phase surface energy to be higher than that of the low-pressure phase, leading to an upshift of the transition pressure (Tolbert and Alivisatos 1995, Machon et al. 2014). This model will be developed in Section 2.1. Using the Landau theory of phase transition, a similar result is obtained by treating the difference of surface energies as a secondary order parameter (see Section 2.2). In addition to the shift of the transition pressure, it has been observed that the pressure range for which both phases (low- and high-pressure phases) coexist is wider at the nanoscale. Such effect can be understood using the extension of the model of Landau, the Ginzburg-Landau theory of phase transitions that accounts for the kinetics and heterogeneity of the system (Section 2.3). Finally, a special phase transformation, size-dependent pressure-induced amorphization, will be treated as a case study. The competition between amorphization and polymorphic transitions will be discussed (Section 2.4).

2.1 Classical Thermodynamics Approach Based on Gibbs Energy

Classical thermodynamics approaches rely on the use of Gibbs energy. When pressure and temperature are control parameters, the Gibbs energy function is expressed as

$$dG = V.dP - S.dT \qquad (1)$$

When G is molar, V is the molar volume (V_m), and S is the molar entropy (S_m).

Considering an isothermal compression, i.e., $-S.dT = 0$, the change of Gibbs energy is then proportional to the molar volume. In a first approximation, this quantity can be considered as constant, i.e., there is a linear relationship between Gibbs energy variation and the increase in pressure (Figure 1).

The surface effect is usually not considered in bulk materials as its contribution can be neglected compared to the volume energy. However, decreasing the particle size modifies the relative contribution of the volume and surface energies. At some point, the energy from the surface cannot be neglected anymore.

Creating a surface dA requires an amount of work equal to $2\gamma dA$, where γ is the surface energy, and the factor two comes from the fact that two surfaces dA are created. From the microscopic point of view, this energy corresponds to the disrupted chemical bonds at the surface.

Including this contribution into the Gibbs energy with A as the control parameter leads to

$$dG = V.dP - SdT + \gamma dA \qquad (2)$$

The factor two is not included because the surface is created during the synthesis (bottom-up approach) and not from the cutting of a bulk material (top-down approach).

In Equation (2), the Gibbs energy is the energy of the system (not molar) as γ has units of $J.m^{-2}$ and dA is in m^2.

If we now consider an isothermal compression of a nanoparticle:

$$dG = V.dP + \gamma dA \qquad (3)$$

In a first approximation, the area does not change with increasing pressure and the surface does not induce additional pressure (meaning that the Laplace pressure is negligible compared to the pressure that will be applied to the nanoparticles). This allows for a direct integration leading to:

$$G = G_0 + V.P + \gamma.A \qquad (4)$$

Considering n moles in the system, the molar Gibbs energy is obtained as:

$$\Delta G_m = V_m.P + \frac{\gamma.A}{n} \qquad (5)$$

As $n = \dfrac{V}{V_m}$,

$$\Delta G_m = V_m.P + \frac{3\gamma.V_m}{R} \qquad (6)$$

for a spherical particle with a radius R.

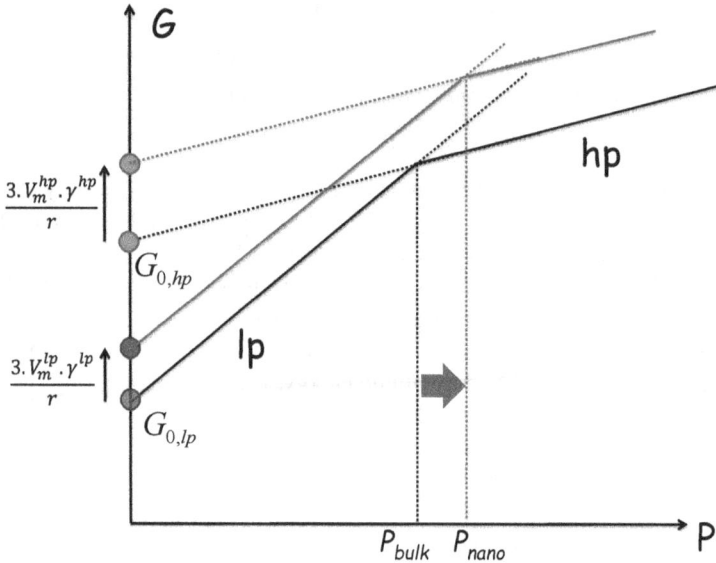

Figure 1. Effect of interfacial energy on phase stability. The phase transition pressure in the bulk material is given by the crossover between the molar Gibbs energy function for each phase (lp for low-pressure and hp for high-pressure phases, respectively). In the case of nanoparticles, the Gibbs energy must be corrected by a surface-related term for each phase (γ^{lp} and γ^{hp}). Usually, the surface tension is higher in the high-pressure phase than in the low-pressure one, which leads to a predicted shift in the transition pressure to a higher value. Reproduced from Ref. (Machon and Mélinon 2015) with permission from the PCCP Owner Societies.

An illustration of these considerations is given in Figure 1. The Gibbs energy is shown as a function of pressure. We consider a transition between a low-pressure phase (lp) and a high-pressure phase (hp) in the cases of:

(i) A bulk material with no size effect. The dark lines show the Gibbs energy of a low-pressure phase (lp) and a high-pressure phase (hp). To a first approximation, the slope of the line is given by the molar volume of the phase (as $dG_m = V_m dP$). The transition pressure is determined by the intercept of the two lines ($G_{lp} = G_{hp}$).

(ii) A nanoparticle of the same composition. Considering $\gamma_{hp} > \gamma_{lp}$ (Tolbert and Alivisatos 1995), where γ_{hp} and γ_{lp} are the surface energies of the high-pressure phase and the low-pressure phase, respectively, we predict an increase in the transition pressure for nanoparticles.

It is important to note that in a realistic case, γ is variable as it corresponds to an interfacial energy term. This term should be taken into account not only because of the surface energy but also because of the presence of defects, interactions with capping molecules, and so on. For instance, this interfacial energy in nanoparticles may vary depending on the synthesis approach (Machon et al. 2011, 2014). As discussed by Carenco et al., in the case of a nanoparticle, one has to consider the different contributions of the surface/interface in a multidimensional plot (defect, capping molecules, etc.) (Carenco et al. 2014). γ results then from a projection on the (G, P) plane and describes the contribution of the different nanoscale-related parameters (such as ligands, shape, size, strain constraint, or the confinement effect).

Using the classical thermodynamics approach, the essential ingredients to understand the variation of the pressure-induced phase transitions as a $1/R$-law are displayed. Alternative thermodynamics models can be used, such as the Landau theory of phase transitions. Using this approach and its extension (Ginzburg-Landau) allows modeling of the heterogeneity associated with the surface state of nanoparticles.

2.2 Landau Theory of Phase Transitions

The Landau theory of phase transitions is based on the definition of an order parameter. This is a physical quantity that is involved in a symmetry-breaking mechanism. By analyzing the symmetries of the parent and the daughter phases, one can determine the order parameter symmetry and deduce its physical nature (Tolédano and Tolédano 1987).

For a long time, it has been believed that the approach developed by Landau is restricted to group-to-subgroup phase transitions. However, Tolédano and Dmitriev demonstrated that the initial Landau theory of phase transitions is an approximation, restricted to small displacements, of a wider formalism (Tolédano and Dmitriev 1996). This extended formalism allows defining an order parameter for any structural phase transition. This is an important starting point as we can assume that an order parameter can always be defined. This gives to this approach the same general character as the Gibbs approach to phase transitions.

Let us first recall the basic approach of the Landau theory of phase transitions using as a starting example of a second-order phase transition. Secondary order parameters will be introduced, and the interface energy will be considered in this description.

For second-order phase transition, the Landau potential is expressed as

$$F(T,P,\eta) = F_0(T,P) + \frac{\alpha}{2}\eta^2 + \frac{\beta}{4}\eta^4 \tag{7}$$

where η is the order parameter, a physical quantity null in one phase and non-null in the other.

Minimization of this energy with respect to the order parameter yields:

$$\frac{\partial F}{\partial \eta} = \eta(\alpha + \beta\eta^2) = 0 \tag{8}$$

which has two solutions

1) Phase I with $\eta = 0$

2) Phase II with $\eta \neq 0$ and

$$\eta^2 = \frac{-\alpha}{\beta} \tag{9}$$

The stability conditions that ensure that the phases are stable are given by $\frac{\partial^2 F}{\partial \eta^2} \geq 0$, which gives the condition $\alpha + 3\beta\eta^2 \geq 0$.

For phase I ($\eta = 0$), the stability condition is then $\alpha \geq 0$.

For phase II ($\eta \neq 0$), this condition is $\alpha + 3\beta\left(\dfrac{-\alpha}{\beta}\right) \geq 0$, leading to $\alpha \leq 0$.

Therefore, to respect the stability conditions in both phases, α should change its sign at the transition. The simplest way of expressing this requirement is to state $\alpha = \alpha_0(T - T_C)$ in the case of temperature-induced phase transitions or $\alpha = \alpha_0(P_c - P)$, in the case of pressure-induced phase transitions.

For instance, let us consider the classic example of the para-to-ferromagnetic phase transition that is the textbook example for introducing the Landau theory of phase transitions. The high-temperature phase is paramagnetic, i.e., it has a magnetization $M = 0$. Below a critical temperature (T_C, Curie temperature), the phase becomes ferromagnetic with $M \neq 0$. The physical quantity that changes at the transition is M, which is considered as the order parameter. It is found that M varies as $(T - T_C)^{1/2}$ by introducing the temperature dependence of α in the expression of η (Equation 9), which is in very good agreement with the experimental finding.

In addition to the primary order parameter that is related to the symmetry-breaking mechanism, secondary order parameters can be introduced. They are not a driving force in the transition mechanism but are coupled physical quantities affected through the phase transformation. The effect of secondary parameters does not affect the symmetry or the nature of the phases involved in the transition. However, they have an effect on the position of the transition line. This is the case of the magnetostriction effect that results from a coupling between the magnetic lattice and the atomic lattice, i.e., the structure. The magnetic transition induced some deformation ε compatible with the new point group and then contributes to the potential.

The Landau potential, including this coupling, is then expressed as:

$$F(T, P, M, \varepsilon) = F_0 + \frac{\alpha}{2}M^2 + \frac{\beta}{4}M^4 + \frac{\delta}{2}M^2\varepsilon \qquad (10)$$

The coupling between the primary order parameter (magnetization M) and the secondary order-parameter (strain ε) depends on the symmetry of the phases. In this case, we use a linear-quadratic coupling that is similar to the case of iron (Tolédano 2010). δ is the coupling constant between the magnetization and strain.

The expression of the potential can be modified:

$$F(T, P, M, \varepsilon) = F_0 + \frac{\alpha^*}{2}M^2 + \frac{\beta}{4}M^4 \qquad (11)$$

with $\alpha^* = \alpha_0(T - T_C^*)$

where $T_C^* = T_C - \dfrac{\delta\varepsilon}{\alpha_0}$

This indicates that the appearance of strain during the transformation induced a shift in the transition temperature without changing the nature of the transition.

An analogy can be developed by considering the interface energy contribution. In nanoparticles, the pressure-induced phase transitions are similar to those in bulk but shifted according to the $1/R$-law discussed in Section 2.1. This suggests the effect of a secondary order parameter related to the size. Therefore, one can introduce a coupling term between the primary order parameter and the surface energies, considered secondary order parameters. Assuming that the volume difference between the two phases is neglected, the modified Landau potential, including a coupling between primary and secondary order parameters, may be written as:

$$F(T,P,\eta) = F_0(T,P) + \frac{\alpha}{2}\eta^2 + \frac{\beta}{4}\eta^4 + \frac{\delta}{2}\eta^2\frac{\gamma_{hp}-\gamma_{lp}}{V}S \tag{12}$$

where S and V are the surface and the volume of the nanoparticle, respectively. Here, δ is the coupling constant between the primary order parameter and the surface energies.

Then, solving the equation $\partial F/\partial \eta = 0$, with the assumption that $\alpha = \alpha_0(P_c - P)$, leads to

$$P_c^* = P_c + 3\delta\frac{\gamma_{hp}-\gamma_{lp}}{\alpha_0 R} \tag{13}$$

where P_c^* is the pressure at which the transition is observed. One then finds the classical dependence on $1/R$ of the transition pressure at the nanoscale.

If a second-order phase transition is used in the development presented here, the generalization to any phase transition, including reconstructive ones (i.e., without group-to-subgroup relationships), is possible and a order parameter can always be defined (Tolédano and Dmitriev 1996). The mathematical treatment is similar and leads to the Equation (13).

Therefore, the general case of nanometric effect on pressure-induced phase transitions, that is, the shift of the transition line with a transition pressure, scales with $1/R$. However, great care should be taken to define the transition pressure in nanocrystals. In most experiments, the transition pressure is defined either as the appearance of the high-pressure phase or at the disappearance of the low-pressure phase or even at the middle of the transition width. This point is crucial because a concomitant effect of pressure-induced phase transitions in nanosized systems is the broadening of the transition width (pressure range where low- and high-pressure phases coexist). For instance, in the case of ZnO, a $1/R$-dependence of the transition pressure has been proposed (Li et al. 2008), with R the characteristic size of the nanocrystal. However, a closer look at the experimental reports indicates that the same definition of the transition pressure was not used (Figure 2). Taking the transition width into account makes the interpretation more complex. No clear size-dependence (as a $1/R$ law) can be drawn. Even though below 15 nm, the transition occurs at a higher pressure than for the bulk, it should be noted that these nanoparticles (obtained by ball-milling, for instance) can also be more defective. A more recent high-pressure study using the same experimental protocol was performed on ZnO nanoparticles of similar size obtained by different synthesis paths (Machon et al. 2014). The main conclusions are that transition features (transition start and end

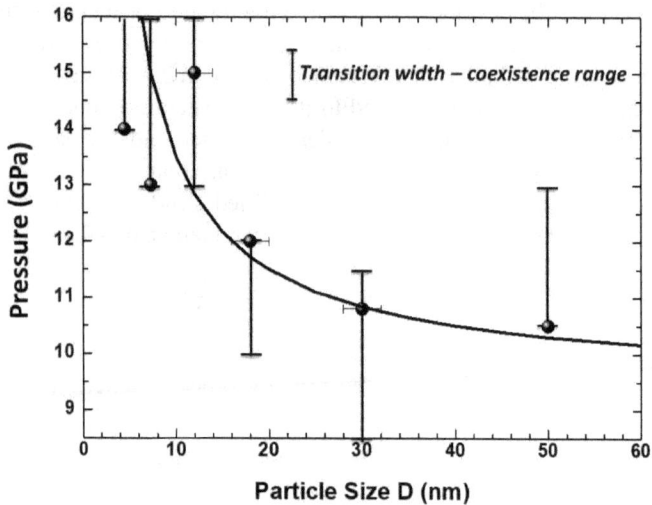

Figure 2. Transition pressures reported in the literature (points) for ZnO nanoparticles as a function of their characteristic size. An apparent $1/R$-law fits the experimental data. However, when including the transition width as error bars—defined from the appearance of the high-pressure phase to the disappearance of the low-pressure phase—a more complex picture rises. Adapted from (Machon et al. 2018b).

pressures, transition width, high-pressure phases) differ in all cases and are mainly related to the presence of defects.

Therefore, it is difficult to extract a simple size-effect when aggregating several experimental studies. Some precautions are required concerning the sample characterization, and an experimental protocol must be followed during high-pressure investigations of nanoparticles in order to obtain a coherent, reliable set of data that can be used by the scientific community (Machon et al. 2014).

2.3 Ginzburg-Landau Theory of Phase Transitions

Another observation regarding the difference between pressure-induced transformations in bulk and nanomaterials is the coexistence range, i.e., the pressure range for which the two phases coexist. It appears that this range is extended when nanoparticles are considered. To discuss such an effect, one can use the preceding model to take into account the heterogeneity of the nanoparticles associated with their surface states. It is recognized that the presence of defects leads to a smearing of the critical thermodynamics' properties. Highly sophisticated models have been proposed to describe these effects. In the following, simplified theoretical models are proposed to give the essential ingredients that allow for a satisfactory understanding of the impact of the defects, such as point defects or surface defects (interface), on pressure-induced phase transitions.

The recurrent criticism of the Landau theory of phase transitions is that it is a mean-field theory. Therefore, it may be hardly applied to heterogeneous materials such as nanoparticles, where the surface plays an important role in leading to a spatial dependency of the order parameter. To resolve such an issue, the Ginzburg-Landau approach includes a spatial dependence of the order parameter using a term with the gradient of the order parameter in the Landau potential. Such a treatment was applied to the case of type-II superconductivity (Ginzburg and Landau 1950). Later, it was used to describe the incommensurate phase (Dzialoshinskii 1964), which shows some loss of the translational symmetry, and, ultimately, to describe amorphization processes (Tolédano and Bismayer 2005, Tolédano 2007). This is also a classical approach used to consider the effect of domain walls. More generally, deviation from the perfect crystallinity and the presence of heterogeneities inside the system may be treated by the following potential:

$$F = F_0 + \frac{\alpha}{2}\eta^2 + \frac{\beta}{4}\eta^4 + \delta\eta^2 \frac{\gamma_{hp} - \gamma_{lp}}{V} S + K(\nabla\eta)^2 \tag{14}$$

The Ginzburg-Landau potential is conservative as long as we use a single order parameter that is identical for all the nanoparticles (narrow size distribution). The additional term $K(\nabla\eta)^2$, the so-called Ginzburg term, corresponds to a kinetic contribution.

The state of minimal energy can be related to the Euler equation after rescaling (Borzi et al. 2005)

$$\frac{\partial F^*}{\partial \eta} = \Delta\eta + \frac{1}{\theta^2}\eta(1-|\eta^2|) = 0 \tag{15}$$

that can be solved analytically as detailed in (Machon et al. 2015). θ corresponds to the width of the transition region.

The nontrivial solution in one dimension leading to the expression of the order parameter is (Borzi et al. 2005)

$$\eta(x) = \frac{e^{\left(\frac{\sqrt{2}}{\theta}\right)x} - 1}{e^{\left(\frac{\sqrt{2}}{\theta}\right)x} + 1} \tag{16}$$

This function is plotted in Figure 3. In the absence of the Ginzburg term ($K = 0$), $\theta \sim 0$, the function behaves like the Heaviside function with a clear-cut upstroke at the transition $P = P_c^*$ (here P_c^* is defined in Equation 13). When K is non-zero, the order parameter is influenced by the "kinetic part". This leads to a spreading order parameter depending on the magnitude of K. This explains why a clear-cut upstroke at the transition is not observed. Increasing the K factor (by considering all the sources of inhomogeneities) introduces a domain where the transition is not clearly defined. This is the origin of the hysteresis, the upstroke-downstroke width being about 4θ. In an ideal case (a single nanoparticle in an ideal medium), the interface between the two phases during the nucleation process is the first origin of phase coexistence.

2.4 Size-dependent Amorphization and Competition between Polymorphic Transformation and Amorphization

Metastable states of matter, such as amorphous states, are controlled by the kinetics of phase transitions. Accessing such states requires energizing processes

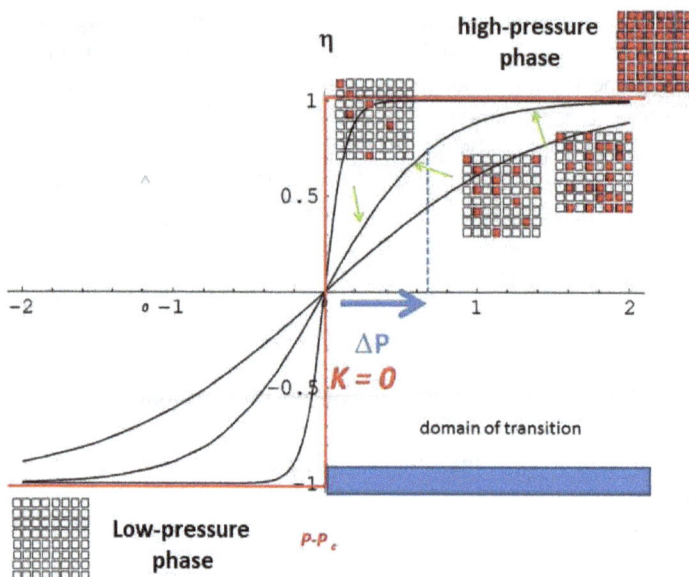

Figure 3. Evolution of the order parameter as a function of pressure. The red curve corresponds to $K = 0$ in the Ginzburg-Landau potential, meaning the absence of order parameter gradient. The bar at the bottom of the graph is the domain of coexistence between the two phases. If the pressure is defined at the inflection point, ΔP is the apparent pressure for the transition. In mono domains (defect free), the window is directly proportional to the gradient. In particles with defects, the window depends on the number of defects. Reprinted (adapted) with permission from (Machon et al. 2015). Copyright (2015) American Chemical Society.

such as irradiation, pressurization, ball-milling, and their combination. In the case of nanoparticles, the interface increases the total energy of the system, and, consequently, nanomaterials show a greater propensity for solid-state amorphization. Thus, regardless of the type of the amorphization process (pressure-, radiation-, or mechanically-induced), the underlying thermodynamic explanation may be identical.

In several cases, pressure-induced amorphization in nanocrystals has been reported. Thus, below a critical particle size, some pressurized nanometric compounds (e.g., TiO_2, Y_2O_3, PbTe) undergo a crystal-to-amorphous transformation instead of a polymorphic transition. For instance, it has been shown that, below a diameter of 10 nm, TiO_2 nanoparticles undergo pressure-induced amorphization (PIA), whereas nano-anatase transforms into a baddeleyite structure when the diameter exceeds 10 nm (Pischedda et al. 2006, Swamy et al. 2006). Such size-dependent PIA has also been reported in Y_2O_3 nanoparticles (Piot et al. 2013). When the particle diameter exceeds 21 nm, the initial cubic structure exhibits a transition to a hexagonal structural phase at 14 GPa. A different scenario is observed for particles with a diameter of 16 nm, for which the cubic phase transforms into an amorphous state at 25 GPa (Wang et al. 2010). A similar effect has been observed in PbTe nanoparticles. Below a diameter of 9 nm, PbTe nanoparticles undergo PIA whereas, with larger particles, a polymorphic transition is observed at 8 GPa (Quan et al. 2013).

The approach based on the Gibbs energy explicitly provides the reasons for amorphization. This transition occurs when the energizing processes (defects, interfacial and elastic energies) destabilize the crystalline phase in favor of the amorphous state, i.e., when the Gibbs energy of the crystalline phase pressure exceeds that of the amorphous state (Machon and Mélinon 2015). The amorphization pressure decreases when the free energy at ambient pressure is high, i.e., when the interfacial energy and the defect concentration are important. This amorphization pressure can be relatively low compared to the polymorphic transition pressure, and then the system is far from equilibrium. However, the difference between the amorphization pressure and the polymorphic transformation pressure is not straightforward to quantify, as the free energies of the amorphous states are not known.

The competition between these two transformation processes (amorphization *vs.* polymorphic) may be discussed using the formalism based on the Ginzburg-Landau theory of phase transitions detailed in the preceding section. As discussed, when the spatial inhomogeneity is limited, the polymorphic transition occurs. In that case, the kinetics term (K) induces a broadening of the transition region.

The amorphization mechanism has been described using a similar approach (Tolédano and Bismayer 2005). However, for amorphization, contrary to polymorphic transitions for which the order parameter is associated with a single critical wave vector, the order parameter associated with the crystal–amorphous transition varies continuously from one point to another. In this description of the crystalline to amorphous transformation under irradiation, the Ginzburg–Landau approach was used to derive the radius of the amorphous region, r_n given by the following relation:

$$r_n = \left(\frac{K}{\alpha_0}\right)^{\frac{1}{2}} (C_c - C_N)^{-\frac{1}{2}} + r_0 \qquad (17)$$

where α_0 is a parameter of the Landau potential, K defines the Ginzburg term in Equation (14), and r_0 is the initial radius of the amorphous embryo. C_N is the defect concentration at which the amorphous embryo nucleates, and C_c is a critical concentration at which merging of amorphous embryos occurs.

Therefore, two size effects can be deduced. First, as in bulk material, when C_N approaches C_c, r_n diverges and leads to amorphization. Such a concentration is explicitly dependent on several nanoparticle parameters, i.e., its size (R) and the interfacial energy (γ) expressed as (Ovid'ko and Sheinerman 2005):

$$C_c = C_0 - \lambda \frac{\gamma}{R} \qquad (18)$$

C_0 is the critical concentration for bulk samples that depends on the materials and the type of defects. Therefore, the radius of the amorphous region is dependent on the nanoparticles' features as, combining (17) and (18):

$$r_n = \left(\frac{K}{\alpha_0}\right)^{\frac{1}{2}} \left[(C_0 - C_N) - \lambda \frac{\gamma}{R}\right]^{-\frac{1}{2}} + r_0 \qquad (19)$$

when the critical defect density is approached, amorphization is favored. Two different parameters may facilitate the amorphization process: (i) reducing the nanoparticle size. As evidenced experimentally, the smaller the nanoparticles, the greater the tendency for amorphization; (ii) increasing the interfacial energy parameter γ by varying the surface state of the nanoparticles.

Therefore, at the transition pressure, two competitive mechanisms must be considered: polymorphism *vs.* amorphization. Three cases can be considered depending on the initial defect concentration C:

(i) $C > C_C$: the nanoparticle is entirely amorphous according to the definition of C_C.

(ii) $C < C_N$: no amorphous nucleates appear, and only polymorphic transformation occurs.

(iii) $C_N < C < C_C$: the final state depends on the comparison between r_n and R. When $r_n > R$, the nanoparticle becomes amorphous; otherwise, amorphous and crystalline states coexist, and the fraction of the amorphous region in the nanoparticle is given by $\left(\dfrac{r_n}{R}\right)^3$.

Considering case (iii) of coexisting crystalline and amorphous regions, kinetic factors need to be contemplated as well. On the one hand, the Ginzburg parameter K is related to the kinetics of a polymorphic transition in nanoparticles as it widens the transition region θ ($\theta \propto K^{1/2}$), as demonstrated in ZnO under pressure (Machon et al. 2014). On the other hand, increasing K leads to a larger amorphous fraction as $r_n \propto K^{1/2}$. Therefore, the kinetics factor tends to slow down the polymorphic transformation, whereas it favors the amorphous state. This statement agrees with the general trend that amorphous states are kinetically favored states.

In the following, we propose to determine the amorphization pressure by adopting a geometrical description of the growth of the amorphous state in the framework of the percolation theory. As the amorphous state becomes energetically favorable and we are far from equilibrium, growth can be considered explosive. This corresponds to the inverse phenomenon known as explosive crystallization leading to a dendritic structure. The associated kinetics being very rapid, the amorphization pressure determined by a geometrical description should be close to the one given by thermodynamic arguments. Therefore, we propose to examine the "growth"/ "propagation" of the amorphous state in a crystalline matrix in greater detail in the framework of invasive percolation theory, as it has been done in the case of radiation-induced amorphization (Salje et al. 1999, Trachenko et al. 2000).

Let us describe the basic assumptions of the model. A Voronoi cell can be defined around each defect. When approaching the stability field of the amorphous state, this cell can present two possible states: crystal-like or amorphous-like. This is a percolating system where we can define the "experimental" crystal-to-amorphous transition at the percolation threshold. Percolating systems have a parameter q that controls the occupancy of sites (Voronoi cell) in the system. This parameter represents the probability of a site being amorphous and increases with increasing the pressure. The Voronoi cells are independent, but an amorphous cell grows during pressurization above the critical pressure. In this case, this corresponds to an invasive percolating system where the amorphous state is the invader and can penetrate an isolated crystallized region (so-called defender). This is called invasive percolation without trapping. It is well established that percolation without trapping, and regular percolation are of the same universality class (Porto et al. 1997). Thus, a microscopic description of the transition at a local level is not necessary. The key point of the description is the fractal structure of the infinite cluster defined just above the percolation threshold. The use of a fractal dimension is dictated by the inhomogeneity of the system, contrary to homogeneous structures for which Euclidean space is adapted. At a critical value $q = q_c$, where the percolation transition takes place, the mean cluster size given by the correlation length ξ diverges following a power law:

$$\xi \approx |q - q_c|^{-\mu} \tag{20}$$

where μ is the critical exponent, which depends on the type of percolation and the type of lattice. The crystal-to-amorphous transition occurs at the percolation threshold. The offset between the pressures of polymorphic transition and of amorphization $\Delta P = P_c^* - P_c^{am}$ c can be determined by (Milovanov and Rasmussen 2005):

$$\Delta P = \frac{\xi^{2\mu-2}}{\alpha_0^{\xi 2\mu} \Gamma 2(3-\mu)} \tag{21}$$

where Γ is the Euler gamma function, ξ is the microscopic correlation length, and α_0 is a numerical parameter. The maximum offset is given for an invasive percolating system $D_f = 2.5$ and yields:

$$\Delta P_{max} \approx \frac{16\xi^{-2}}{9\alpha_0 \pi} \tag{22}$$

The microscopic correlation length ξ is defined in percolation theory as the typical radius of the largest finite cluster. In the case of nanoparticles, ξ cannot exceed R, the radius of the nanoparticle. This spatial limitation leads to:

$$P_c^{am} = P_c^* - \frac{16}{9\pi\alpha_0 R^2} \tag{23}$$

where P_c^* is the polymorphic transition pressure in the nanoparticle (Equation 13). Finally,

$$P_c^{am} = P_c + 3\delta \frac{\gamma_{hp} - \gamma_{lp}}{\alpha_0 R} - \frac{16}{9\pi\alpha_0 R^2} \tag{24}$$

where P_c corresponds to the polymorphic transition pressure in the bulk material.

Using these results, one can draw a schematic pressure-size phase diagram showing two different regimes: polymorphic transition *vs.* pressure-induced amorphization (Figure 4).

It has been demonstrated that amorphization occurs only when a sufficient defect density is attained in the nanoparticles (Machon et al. 2011, Piot et al. 2013), i.e., when the number of sites is sufficient to reach the percolation threshold. For typical

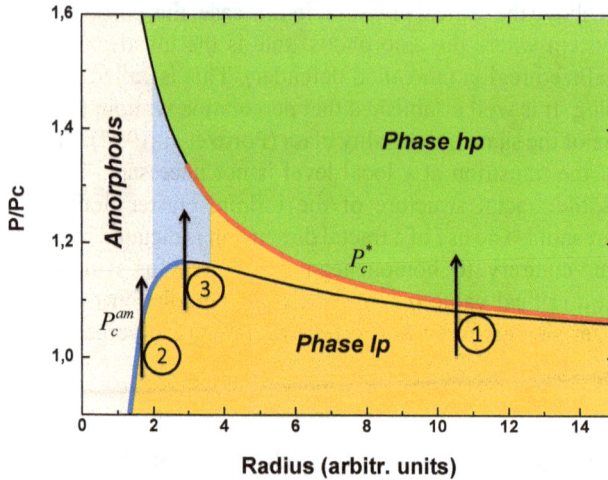

Figure 4. Schematic phase diagram showing different regimes. For large particles, the low-pressure phase (lp) transforms into a high-pressure phase (hp) at a higher pressure (P_c^*) than in the bulk material when the radius decreases (red line). Once a critical size (depending on the material, the surface state, and the defect density) is reached, there exists a cross-over between this polymorphic transition and pressure-induced amorphization (P_c^{am}). In this case, the amorphization pressure decreases with decreasing size (blue line). Reproduced from Ref. (Machon and Mélinon 2015) with permission from the PCCP Owner Societies.

large nanoparticles, no amorphization is observed. In this case, the transition pressure follows a $1/R$-dependence (red curve – arrow 1 in Figure 4) as determined by the classical thermodynamics approach or by the Landau theory of phase transitions (see Sections 2.1 and 2.2). When the defect density is sufficient, then amorphization may occur at a critical size (which depends on the materials: ~ 10 nm in TiO_2 (Pischedda et al. 2006, Swamy et al. 2006), ~ 9 nm in PbTe (Quan et al. 2013), ~ 16 nm in Y_2O_3 (Wang et al. 2010)). In that case, a crossover can occur, and amorphization becomes the dominant mechanism (blue zone in Figure 4). One striking consequence is that $\Delta P < 0$, i.e., the crystal-to-amorphous transition occurs at a lower pressure than the polymorphic transformation (Figure 4 – arrow 2). This prediction has been observed in all experimental reports of size-dependent PIA.

3. Probing Mechanical Properties of Nanomaterials

Using high-pressure tools on nanomaterials is also an interesting way to study the mechanics of nanomaterials (nanomechanics) when usual techniques are not relevant at this nanoscale. For instance, the mechanics of carbon nanotubes have been investigated thanks to high-pressure experiments, allowing to validate theoretical models (Caillier et al. 2008, Torres-Dias et al. 2017). In the following, we will illustrate this methodology through the examples of graphene (and, more generally, 2D systems) and graphene-based nanocomposites.

3.1 Graphene

A new family of nanomaterials has come recently into the scene: 2D materials. Because of their structure, the surface is predominant, and the majority of atoms - if not all - are in interaction with their surroundings. In graphene, all atoms are surface atoms, whereas in transition metal dichalcogenides or in tri-layered graphene, a layer of atoms is intercalated between two surfaces. 2D materials are usually grown on surfaces or deposited on a substrate through different methods.

2D systems exhibit an outstanding stretchability, i.e., the ability to sustain high strain rates without failure, which opens new opportunities unattainable for bulk materials. Actually, the accessible strains in 2D systems are one order of magnitude higher than in 3D materials. This allows anticipating huge effects on physical properties. However, if the strain transmission in 3D materials is well understood using the continuum mechanics laws, these laws fail for 2D materials. Let us consider the case of pressure, which is a force divided by a surface: how can pressure be defined and transmitted laterally on a sample that shows an atomic thickness and for which the concept of surface is irrelevant? To explain such a seeming paradox, it is reasonable to assume that the stresses are transmitted through the substrate deformation. How the strain is transferred between the substrate and the deposited 2D material is a challenging question that is important to answer before using stress as a control parameter of the 2D physical properties.

High-pressure methods constitute a privileged technique to probe the mechanical properties of the 2D system, graphene in particular (Sun et al. 2021). Several experimental works report the effect of high pressure on Raman or photoluminescence spectra for several 2D materials (Nicolle et al. 2011, Machon et al. 2018a, Carrascoso et al. 2021, Proctor et al. 2009, Filintoglou et al. 2013, Nayak et al. 2015, Alencar et al. 2017, Bousige et al. 2017). In most high-pressure studies, physical properties (phonons, bandgap, etc.) versus pressure are represented as if pressure was totally transferred to the 2D samples. However, several works have shown that this is not the case and that substrate compression plays a fundamental role. This is well evidenced by the different pressure dependence of the Raman peaks as a function of the substrate on which the 2D material (graphene in this case) is deposited (Bousige et al. 2017).

Instead of assuming a total transfer of the applied pressure, one can formulate another hypothesis, based on the fact that the G-mode of graphene measured by Raman spectroscopy is the signature of the C-C stretching vibration. Therefore, its pressure-induced evolution should not depend on the substrate if pressure is totally transmitted. To explain such dependence, one could assume that the strain is (partially) transmitted from the substrate (deformed by pressure) to the 2D materials leading to $\varepsilon_{2D} = \alpha.\varepsilon_{substrate}$ where ε_i is the strain in the 2D sample and in the substrate, respectively, and α is the strain transfer function. Considering the bond scale, one considers the linear bulk modulus of both the 2D materials (β_{2D}) and the substrate (β_S), which leads to $\alpha \dfrac{P}{\beta_S} = \dfrac{\sigma}{\beta_{2D}}$, where σ is the stress transmitted by the deformation of the substrate.

This point is important as it appears that, in the case of strain transmission, the stress experienced by the graphene sample is not equal to the applied pressure. It implies that plotting the Raman shift as a function of pressure is not adequate. The relevant parameter is the induced stress σ expressed as

$$\sigma = \alpha \left(\frac{\beta_{2D}}{\beta_S} \right) P \qquad (25)$$

The value of α depends on the nature of the substrate and is more important for substrates having lower compressibility (a higher bulk modulus). A consequence is the impossibility - energetically speaking - of graphene to attain the compression imposed by the substrate, which will lead to the realization of a buckling field.

The pressure dependence of the G-mode of graphene not only depends on the nature of the substrate but also on the number of layers. On a Si/SiO_2 substrate, in the case of monolayer and bilayer graphene, the slope $\partial\omega_G / \partial P$ is larger than the one obtained for trilayers and multilayers (Figure 5). Typically, the slope is around 8 $cm^{-1} \cdot GPa^{-1}$ for graphene and bilayer, whereas it is close to the value found in graphite (~ 4.4 $cm^{-1} \cdot GPa^{-1}$) for trilayers and further (Nicolle et al. 2011). Such observations may be explained considering the graphene systems as a membrane in adhesion with the substrate in which its bending rigidity is modified by increasing the number of graphene layers. A calculation considering the competition between the adhesion energy (gain) and bending energy (cost) shows that the unbinding event between a graphene stacking and a SiO_2 substrate occurs for a number of layers higher than two (Nicolle et al. 2011). Therefore, the strain cannot be transmitted anymore through the substrate deformation, and for trilayer graphene, the pressure is

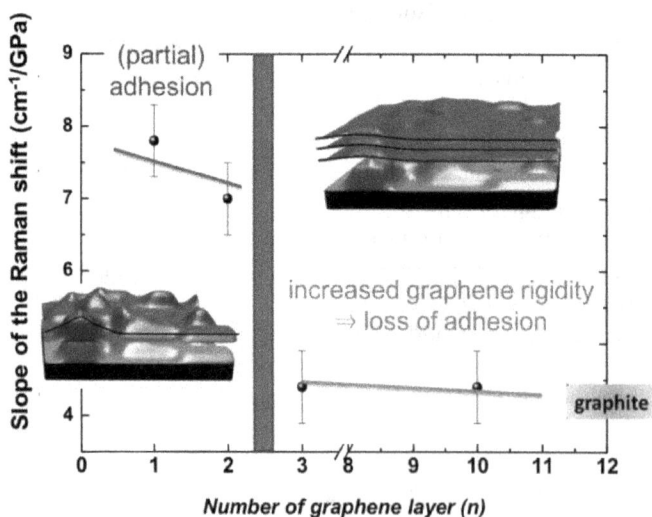

Figure 5. Pressure-dependence of the G-peak as a function of the number of graphene layers (n). For $n = 1$ (single layer) and $n = 2$ (bilayer), the slope is significantly higher than for samples with $n \geq 3$. The regime change for $n > 2$ is interpreted as a loss of adhesion related to the higher rigidity of the multilayer system as depicted by the images. Adapted from (Nicolle et al. 2011).

then directly transmitted to the sample from the pressure transmitting medium, and the pressure dependence of the G-band is similar to graphite.

3.2 *Probing the Coupling between the Components of a Nanocomposite using High-Pressure Raman Spectroscopy*

With the advent of graphene synthesis, an avenue in nanocomposite science and technologies has emerged with the opportunity to modify the surface chemistry of materials by coating the surface with graphene (Kinloch et al. 2018, Sauze et al. 2020). This opened up the field of graphene-based nanocomposites with applications in energy and environmental areas, among others (Huang et al. 2012, Stankovich et al. 2006, Chang and Wu 2013, Papageorgiou et al. 2017). If the chemical and physical properties of a nanocomposite highly depend on the morphology of the different component materials, the nature of their interface is also a major determining factor in the resulting properties (Anagnostopoulos et al. 2015, Dai et al. 2019).

In a recent study, a graphene-mesoporous germanium nanocomposite (Gr-MP-Ge) was studied under pressure using Raman spectroscopy (Machon et al. 2021). Two frequency regions have received a special interest: (i) the region [20–500 cm^{-1}] that allows monitoring of the pressure-induced transformations of the germanium substrate; (ii) the region [1500–1850 cm^{-1}] corresponding to the G-peak of the graphene-like coating.

The pressure-induced variations of both Raman signatures were found to be related. First, the graphene G-peak pressure-dependency is larger than that expected for graphene-based materials. This effect recalls what is observed in graphene deposited on substrates. This pressure-dependence has been explained based on the additional stress component generated by the substrate deformation in the previous section. During compression, the mesoporous germanium substrate undergoes a pressure-induced transition around 9 GPa with a volume reduction of $\sim$ 20%, which leads to a decoupling of interactions between the two components of the nanocomposite. Above this pressure, the pressure-induced Raman shift of the G-peak returns to that of free-standing graphite (Figure 6).

In the first pressure range, from 0.1 GPa to ~7 GPa, the G-peak position shifts linearly with pressure by 5.8(1) cm^{-1}·GPa^{-1}. Above 8 GPa, the peak variation deviates from this linear trend. After a transitional range between ~8 GPa and 13.5 GPa, a linear variation is found again but with a different linear pressure coefficient of 4.5(2) cm^{-1}·GPa^{-1}. The same slope is observed on the decompression path. This value of the slope for the G-peak position as a function of pressure is similar to that of free-standing graphite and lies between 4.2 and 4.7 cm^{-1}·GPa^{-1}. The change in the pressure-induced shift regime is concomitant with the pressure-induced phase transition of the germanium substrate.

Thanks to Equation 25, the value of α can be determined using the bulk modulus of cubic germanium (β_s = 3 where B_0 = 75 GPa is the bulk modulus of the cubic diamond Ge phase) (Young 1991). The value of β_{2D} is taken using the in-plane linear bulk modulus of graphite, if the reduction of dimensionality does not induce

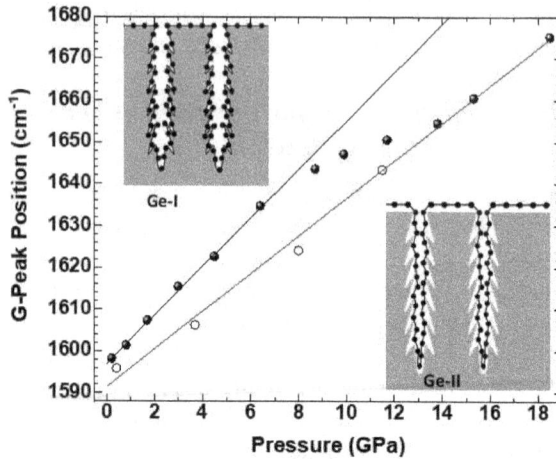

Figure 6. Peak position of the G-peak in the Gr-MP-Ge as a function of pressure on compression (full circles) and on decompression (open circles). The lines show the fits of the G-peak position before and after the pressure-induced transition of the Ge. The slope of the line before the transition is is 5.8(1) cm^{-1}·GPa^{-1} and that of the line after the transitoin is 4.5(2) cm^{-1}·GPa^{-1}. Adapted from (Machon et al. 2021).

modifications of the C–C chemical bond and the associated mechanical properties. Thanks to X-ray diffraction experiments on graphite, it was determined that β_{2D} = 1250GPa (Hanfland 1989).

The calculated value of ~ 23% indicates a rather strong coupling between the carbon layers and the mesoporous Ge structure. For comparison, this parameter is 20% in the case of a monolayer of exfoliated graphene deposited on a Si/SiO$_2$ substrate, which is twice more compressible in planar geometry (Bousige et al. 2017).

Therefore, pressure-induced variations of the Raman spectral features are a valuable tool for exploring the coupling between the components of a graphene-based nanocomposite. It allows quantifying the interactions between these components and observing the unbinding of the components. In addition, the pressure cycling allows modifying the nature of one component, here, from crystal to amorphous structure in mesoporous germanium, leading to the opportunity of tuning the associated physical properties such as thermal or electrical conductivities. This is an important step in the design of nanomaterials (Machon et al. 2018b). Finally, this work proposes a unified protocol that allows for characterizing and comparing the interactions between graphene and the other component in different graphene-based nanocomposites. The results of this study invite us to systematically explore graphene-based (nano) composite materials under pressure.

4. Conclusions

In this chapter, we illustrated the valuable insights of studying nanomaterials under high pressure. First, variations in pressure-induced transitions allow for a better understanding of the thermodynamics at the nanoscale. The effect of the size, and more generally, that of the interface can be better apprehended with the development of appropriate models. The different size effects are phenomenologically described. The size-dependent pressure-induced amorphization is described using a geometrical model. It allows reproducing the fact that this amorphization occurs at a lower pressure than an underlying polymorphic transition. In the second part, we illustrate how high-pressure experiments on nanomaterials permit a better understanding of the mechanics of nanostructure (nanomechanics). It is shown that continuum mechanics is applicable but must be used with care due to additional effects related to the size and surface. The *a priori* assumption that pressure is directly transmitted must be questioned for correctly interpreting the pressure-induced shifts of the Raman peaks in 2D materials. Considering strain transmission between the substrate and the 2D material allows us to understand the different pressure dependencies of the Raman shifts that vary with the substrate. In the case of graphene, the regime change between strain transmission and stress transmission occurs for a number of layers larger than two. This is related to the delamination of the few-layer sample from the substrate. The strain transmission may vary depending on the substrate and its surface state. A strain transfer function can be defined and reflects the degree of coupling between the graphene and its substrate. Therefore, using high-pressure experiments allows for quantifying the coupling between components of a graphene-based nanocomposite.

Acknowledgements

The author warmly thanks Alfonso San Miguel, Vittoria Pischedda, Sylvie Le Floch, and Patrice Mélinon from Université de Lyon 1 for the collaboration for several years on the subjects presented in this work.

References

Alencar, R.S., Saboia, K.D.A., Machon, D., Montagnac, G., Meunier, V., Ferreira, O.P., San-Miguel, A. and Souza Filho, A.G. (2017). Atomic-layered MoS$_2$ on SiO$_2$ under high pressure: Bimodal adhesion and biaxial strain effects. *Phys. Rev. Mater.*, 1: 024002.

Anagnostopoulos, G., Androulidakis, C., Koukaras, E.N., Tsoukleri, G., Polyzos, I., Parthenios, J., Papagelis, K. and Galiotis, C. (2015). Stress transfer mechanisms at the submicron level for graphene/polymer systems. *ACS Appl. Mater. Interfaces*, 7: 4216.

Bai, F., Bian, K., Huang, X., Wang, Z. and Fan, H. (2019). Pressure induced nanoparticle phase behavior, property, and applications. *Chem. Rev.*, 119: 7673–7717.

Borzi, A., Grossauer, H. and Scherzer, O. (2005). Analysis of iterative methods for solving a ginzburg-landau equation. *Int. J. Comput. Vision*, 64: 203–219.

Bousige, C., Balima, F., Machon, D., Pinheiro, G.S., Torres-Dias, A., Nicolle, J., Kalita, D., Bendiab, N., Marty, L., Bouchiat, V., Montagnac, G., Souza Filho, A.G., Poncharal, P. and San-Miguel, A. (2017). Biaxial strain transfer in supported graphene. *Nano Lett.*, 17: 21.

Caillier, Ch., Machon, D., San-Miguel, A., Arenal, R., Montagnac, G., Cardon, H., Kalbac, M., Zukalova, M. and Kavan, L. (2008). Probing high-pressure properties of single-wall carbon nanotubes through fullerene encapsulation. *Phys. Rev. B*, 77: 125418.

Carenco, S., Portehault, D., Boissière, C., Mézailles, N. and Sanchez, C. (2014). 25th anniversary article: Exploring nanoscaled matter from speciation to phase diagrams: Metal phosphide nanoparticles as a case of study. *Adv. Mater.*, 26: 371–390.

Carrascoso, F., Li, H., Frisenda, R. and Castellanos-Gomez, A. (2021). Strain engineering in single-, bi- and tri-layer MoS_2, $MoSe_2$, WS_2 and WSe_2. *Nano Research*, 14: 1698–1703.

Chang, H. and Wu, H. (2013). Graphene-based nanocomposites: Preparation, functionalization, and energy and environmental applications. *Energy Environ. Sci.*, 6: 3483.

Chen, X. and Mao, S.S. (2007). Titanium dioxide nanomaterials: Synthesis, properties, modifications, and applications. *Chem. Rev.*, 107: 2891.

Dai, Z., Liu, L. and Zhang, Z. (2019). Strain Engineering of 2D materials: Issues and opportunities at the interface. *Adv. Mater.*, 31: 1805417.

Diebold, U. (2003). The surface science of Titanium dioxide. *Surf. Sci. Rep.*, 48: 53–229.

Dzialoshinskii, I.E. (1964). *Sov. Phys. JETP*, 19: 960.

Filintoglou, K., Papadopoulos, N., Arvanitidis, J., Christofilos, D., Frank, O., Kalbac, M., Parthenios, J., Kalosakas, G., Galiotis, C. and Papagelis, K. (2013). Raman spectroscopy of graphene at high pressure: Effects of the substrate and the pressure transmitting media. *Phys. Rev. B*, 88: 045418.

Ginzburg, V.L. and Landau, L.D. (1950). On the theory of superconductivity. *Sov. Phys. JETP*, 20: 1064–1082.

Girão, H.T., Cornier, T., Daniele, S., Debord, R., Caravaca, M.A., Casali, R.A., Mélinon, P. and Machon, D. (2017). Pressure-induced disordering in SnO_2 nanoparticles. *J. Phys. Chem. C*, 121: 15463–15471.

Hanfland, M., Beister, H. and Syassen, K. (1989). Graphite under pressure: Equation of state and first-order Raman modes. *Phys. Rev. B*, 39: 12598.

Huang, X., Qi, X., Boey, F. and Zhang, H. (2012). Graphene-based composites. *Chem. Soc. Rev.*, 41: 666–686.

Illy, S., Tillement, O., Machiz, F., Dubois, J.M., Massicot, F., Fort, Y. and Ghanbaja, J. (1999). First direct evidence of size-dependent structural transition in nanosized nickel particles. *Philosphical Magazine A*, 79: 1021–1031.

Jayaraman, A. (1986). Ultrahigh pressures. *Rev. Sci. Instrum.*, 57: 1013.

Kinloch, I.A., Suhr, J., Lou, J., Young, R.J. and Ajayan, P.M. (2018). Composites with carbon nanotubes and graphene: An outlook. *Science*, 362: 547–553.

Li, S., Wen, Z. and Jiang, Q. (2008). Pressure-induced phase transition of CdSe and ZnO nanocrystals *Scr. Mater.*, 59: 526.

Machon, D. and Mélinon, P. (2015). Size-dependent pressure-induced amorphization: A thermodynamic panorama. *Phys. Chem. Chem. Phys.*, 17: 903.

Machon, D., Bousige, C., Alencar, R., Torres-Dias, A., Balima, F., Nicolle, J., de Sousa Pinheiro, G., Souza Filho, A.G. and San-Miguel, A. (2018a). Raman scattering studies of graphene under high pressure. *J. Raman Spectrosc.*, 49: 121.

Machon, D., Daniel, M., Bouvier, P., Daniele, S., Le Floch, S., Mélinon, P. and Pischedda, V. (2011). Interface energy impact on phase transitions: The case of TiO_2 nanoparticles. *J. Phys. Chem. C*, 115: 22286–22291.

Machon, D., Piot, L., Hapiuk, D., Masenelli, B., Demoisson, F., Piolet, R., Ariane, M., Mishra, S., Daniele, S., Hosni, M., Jouini, N., Farhat, S. and Mélinon, P. (2014). Thermodynamics of nanoparticles: Experimental protocol based on a comprehensive Ginzburg-Landau interpretation. *Nano Lett.*, 14: 269–276.

Machon, D., Pischedda, V., Le Floch, S. and San-Miguel, A. (2018b). Perspective: High pressure transformations in nanomaterials and opportunities in material design. *J. Appl. Phys.*, 124: 160902.

Machon, D., Sauze, S., Arés, R. and Boucherif, A. (2021). Probing the coupling between the components in a graphene–mesoporous germanium nanocomposite using high-pressure Raman spectroscopy. *Nanoscale Adv.*, 3: 2577–2584.

Milovanov, A.V. and Rasmussen, J.J. (2005). Fractional generalization of the Ginzburg–Landau equation: an unconventional approach to critical phenomena in complex media. *Phys. Lett. A*, 337: 75.

Navrotsky, A. (2001). Thermochemistry of nanomaterials. *Reviews in Mineralogy and Geochemistry*, 44: 73–103.

Navrotsky, A. (2003). Energetics of nanoparticle oxides: Interplay between surface energy and polymorphism. *Geochem. Trans.*, 4: 34–37.

Nayak, A.P., Pandey, T., Voiry, D., Liu, J., Moran, S.T., Sharma, A., Tan, C., Chen, C.-H., Li, L.-J., Chhowalla, M., Lin, J.-F., Singh, A.K. and Akinwande, D. (2015). Pressure-dependent optical and vibrational properties of monolayer molybdenum disulfide. *Nano Lett.*, 15: 346.

Nicolle, J., Machon, D., Poncharal, P., Pierre-Louis, O. and San-Miguel, A. (2011). Pressure-mediated doping in graphene. *Nano Lett.*, 11: 3564.

Ovid'ko, I.A. and Sheinerman, A.G. (2005). Irradiation-induced amorphization processes in nanocrystalline solids. *Appl. Phys. A: Mater. Sci. Process.*, 81: 1083–1088.

Papageorgiou, D.G., Kinloch, I.A. and Young, R.J. (2017). Mechanical properties of graphene and graphene-based nanocomposites. *Prog. Mater. Science*, 90: 75.

Piot, L., Le Floch, S., Cornier, T., Daniele, S. and Machon, D. (2013). Amorphization in nanoparticles. *J. Phys. Chem. C*, 117: 11133–11140.

Pischedda, V., Hearne, G.R., Dawe, A.M. and Lowther, J.E. (2006). Ultrastability and enhanced stiffness of ~ 6 nm TiO_2 nanoanatase and eventual pressure-induced disorder on the nanometer scale. *Phys. Rev. Lett.*, 96: 035509.

Porto, M., Havlin, S., Schwarzer, S. and Bunde, A. (1997). Optimal path in strong disorder and shortest path in invasion percolation with trapping. *Phys. Rev. Lett.*, 79: 4060.

Proctor, J., Gregoryanz, E., Novoselov, K., Lotya, M., Coleman, J. and Halsall, M. (2009). High-pressure Raman spectroscopy of graphene. *Phys. Rev. B*, 80: 073408.

Quan, Z., Luo, Z., Wang, Y., Xu, H., Wang, C., Wang, Z. and Fang, J. (2013). Pressure-induced switching between amorphization and crystallization in PbTe nanoparticles. *Nano Lett.*, 13: 3729–3735.

Ram, S. (2001). Allotropic phase transformations in HCP, FCC, and BCC metastable structures in Co-nanoparticles. *Materials Science and Engineering A*, 304-306: 923–927.

Roduner, E. (2006). Size matters: Why nanomaterials are different. *Chem. Soc. Rev.*, 35: 583–592.

Salje, E.K.H., Chrosch, J. and Ewing, R.C. (1999). Is "metamictization" of zircon a phase transition? *Am. Mineral.*, 84: 7.

San-Miguel, A. (2006). Nanomaterials under high-pressure. *Chem. Soc. Rev.*, 35: 876–889.

Sauze, S., Aziziyan, M.R., Brault, P., Kolhatkar, G., Ruediger, A., Korinek, A., Machon, D., Arès, R. and Boucherif, A. (2020). Integration of 3D nanographene into mesoporous germanium. *Nanoscale*, 12: 23984–23994.

Saviot, L., Machon, D., Mermet, A., Murray, D.B., Adichtchev, S., Margueritat, J., Demoisson, F., Ariane, M. and Marco de Lucas, M.C. (2012). Quasi-free nanoparticle vibrations in a highly compressed ZrO2 nanopowder. *J. Phys. Chem. C*, 116: 22043–22050.

Shen, G. and Mao, H.K. (2017). High-pressure studies with x-rays using diamond anvil cells. *Rep. Prog. Phys.*, 80: 016101.

Stankovich, S., Dikin, D.A., Dommett, G.H.B., Kohlhaas, K.M., Zimney, E.J., Stach, E.A., Piner, R.D., Nguyen, S.T. and Ruoff, R.S. (2006). Graphene-based composite materials. *Nature*, 442: 282.

Sun, Y.W., Papageorgiou, D.G., Humphreys, C.J., Dunstan, D.J., Puech, P., Proctor, J.E., Bousige, C., Machon, D. and San Miguel, A. (2021). Mechanical properties of graphene. *Appl. Phys. Rev.*, 8: 021310.

Swamy, V., Kuznetsov, A., Dubrovinsky, L.S., McMillan, P.F., Prakapenka, V.B., Shen, G. and Muddle, B.C. (2006). Size-dependent pressure-induced amorphization in nanoscale TiO_2. *Phys. Rev. Lett.*, 96: 135702.

Tolbert, S.H. and Alivisatos, A.P. (1995). High-pressure structural transformations in semiconductor nanocrystals. *Annu. Rev. Phys. Chem.*, 46: 595–626.

Tolédano, P. (2007). Theory of the amorphous solid state: Non-directional elastic vortices and a superhard crystal state. *Europhys. Lett.*, 78: 46003.

Tolédano, P. and Bismayer, U. (2005). Phenomenological theory of the crystalline-to-amorphous phase transition during self-irradiation. *J. Phys.: Condens. Matter.*, 17: 6627–6634.

Tolédano, P. and Dmitriev, V. (1996). *Reconstructive Phase Transitions*. World Scientific, Singapore.

Tolédano, P. and Tolédano, J.C. (1987). *Landau Theory of Phase Transitions: The Application to Structural, Incommensurate, Magnetic And Liquid Crystal Systems*. World Scientific Publishing Company.

Tolédano, P., Katzke, H. and Machon, D. (2010). Symmetry-induced collapse of ferromagnetism at the α–ε phase transition in iron. *J. Phys.: Condens. Matter.*, 22: 466002.

Torres-Dias, A.C., Cerqueira, T.F.T., Cui, W., Marques, M.A.L., Botti, S., Machon, D., Hartmann, M.A., Sun, Y., Dunstan, D.J. and San-Miguel, A. (2017). From mesoscale to nanoscale mechanics in single-wall carbon nanotubes. *Carbon*, 123: 145–150.

Trachenko, K., Dove, M.T. and Salje, E.K.H. (2000). Modelling the percolation-type transition in radiation damage. *J. Appl. Phys.*, 87: 7702–7707.

Wang, L., Yang, W., Ding, Y., Ren, Y., Xiao, S., Liu, B., Sinogeikin, S.V., Meng, Y., Gosztola, D.J., Shen, G., Hemley, R.J., Mao, W.L. and Mao, H.K. (2010). Size-dependent amorphization of nanoscale Y_2O_3 at high pressure. *Phys. Rev. Lett.*, 105: 095701.

Young, D.A. (1991). *Phase Diagrams of the Elements*. Univ. of California Press, Berkley, 1991.

Chapter 9

Future Perspectives
Nanomaterials, Industry, Legislations, and Dreams

Manuel Ahumada[1,2,*] and *María Belén Camarada*[3,4,*]

1. Introduction

Lately, nanotechnology has been recognized as the driving force behind a new industrial revolution. Since the term 'nanotechnology' was first mentioned by Norio Taniguchi in 1974 (Taniguchi 1974), its meaning has changed constantly, reflecting on how science, politics, and media depict the concept. At the beginning of the '90s, the term's popularity increased, identifying the technology as a powerful and visionary tool. Challenging and futuristic projects were proposed, like the space elevator made of CNTs (Edwards 2000) and nanobots roaming the bloodstream to repair tissues or eliminate specifically cancer cells (Saha 2009). The 2000 decade started with the US National Nanotechnology Initiative (NNI), which was significantly influenced by the auspicious and potential applications for the area (Roco and Bainbridge 2005). The economic potential of nanotechnology came to focus with considerable funding from the governments. For the first time, nanotechnology emerged to initiate the next economic boom. In the same decade, a confident forecast made by the US National Science Foundation of a trillion-dollar market for nanotechnology-related products in 2015 and 6 million new job opportunities generated a huge interest

[1] Centro de Nanotecnología Aplicada, Facultad de Ciencias, Ingeniería y Tecnología, Universidad Mayor, Camino La Pirámide 5750, Huechuraba, Santiago, RM.
[2] Escuela de Biotecnología, Facultad de Ciencias, Ingeniería y Tecnología, Universidad Mayor, Camino La Pirámide 5750, Huechuraba, Santiago, RM.
[3] Laboratorio de Materiales Funcionales, Departamento de Química Inorgánica, Facultad de Química y de Farmacia, Pontificia Universidad Católica de Chile, Santiago, 7820436, Chile.
[4] Centro Investigación en Nanotecnología y Materiales Avanzados, CIEN-UC, Pontificia Universidad Católica de Chile, Santiago, Chile.
* Corresponding authors: manuel.ahumada@umayor.cl; mbcamara@uc.cl

from investment companies that set up nanotechnology in the stock indices (Roco 2011). Of course, expectations grew enormously, proposing cancer cure for 2015, robots in nanomedicine for the directed and conducted treatment of injuries and diseases, among others. However, the situation changed dramatically when the first negative news came from civil rights organizations and new studies cataloging carbon nanotubes as the new asbestos (Brown 2002). The toxicologic findings indicated the potential health risks of nanoproducts and allowed social and environmental organizations to expose the conflicts about these new nanoparticle systems. After that, the economic boom narrowed, and nanotechnology was under the scrutiny of legal and medical regulation but kept its development with great interest from the most important economies in the world. Nowadays, it is more apparent that life quality and economic improvement based on nanotechnology cannot be at the expense of human and environmental damage or risks. Therefore, risk assessment and regulations on nanomaterials are required to advance as a society (Figure 1).

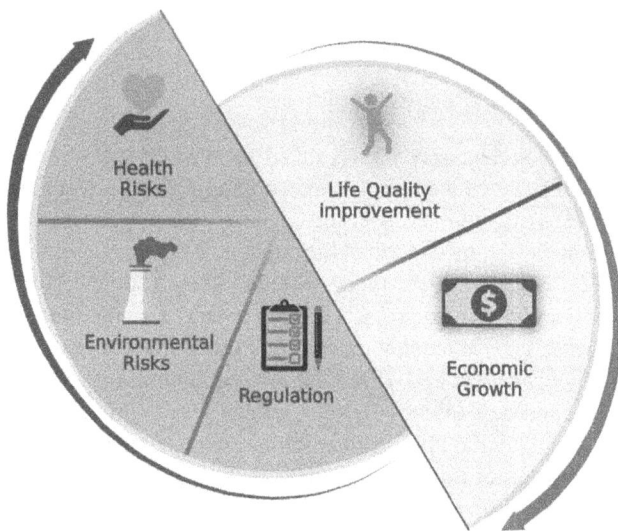

Figure 1. Nanomaterials' impact on society requires a perfect balance between improvements and risk assessments.

2. Markets Evolution

In 2000, nanotechnology raised extremely high expectations and was believed to be the core of the economic revolution, potentially destroying many old companies (Eaglesham 2005, Koshovets and Ganichev 2017). In 2003, the US Government authorized $3.4 billion in federal nanotechnology spending from 2004 through 2008. In 2006, predictions stated that nanotechnology would represent a $2.6 trillion industry by 2014 (Manoharan 2008). During this time, media and scientists created effective information propaganda around this emerging technology to attract potential investors' attention, which led to unrealistic expectations. The exaggerated prospects allowed stakeholders to obtain significant business advantages. Many startups added "nano" to their names to secure funding, and as a result, many of these companies had greatly inflated market cap compared to revenue (Miller et al. 2004). At the same time, some countries developed programs to reach the greatest capabilities in nanotechnology, recognizing its relevance in the industrial field and different social requirements. Besides the NNI in the USA, Japan started programs with fundamental research, nanomaterials, and nanoelectronic devices. Canada opened the National Institute of Nanotechnology and Denmark inaugurated research centers focusing on interdisciplinary nanoscience and nanosystems engineering approaches. The Netherlands considered R&D programs focused on microelectronics and electronic devices, and Poland supported nanotechnology education programs. At the same time, UK started the micro and nanotechnology manufacturing initiative for a new network of micro and nanotechnology facilities (Ezema et al. 2014).

Today, nanotechnology is a priority field for innovation leaders and is a crucial element of advanced manufacturing. Its market now includes producing and applying physical, chemical, and biological devices that use a wide diversity of nanomaterials. Its global market was valued at $1.76 billion in 2020, and the projection is to reach $33.63 billion by 2030, registering a CAGR of 36.4% from 2021 to 2030 (Research 2020).

As can be concluded, nanotechnology has intensively grown over the last years. However, some nanomaterials are related to health and environmental risks that must be considered for responsible and balanced use. Despite the different investigations aimed to clarify and prevent these risks, the estimation of long-term consequences is not yet complete (Klaine et al. 2012). Another complication is the heterogeneity of these compounds to evaluate them as a whole set. In addition, calculating and measuring the potential impacts and risks involves factors from different fields, requiring experts from diverse scientific areas like physicists, engineers, chemists, biologists, and toxicologists. Considering nanomaterials' relevance and their extensive use, there have been limited efforts and investments to study their potential risks and environmental consequences. Therefore, focusing on these specific aspects is necessary to avoid adverse effects on ecosystems and human health (Sengul and Asmatulu 2020).

3. Health and Environmental Risks

As the properties of nanomaterials are distant from bulk materials, a risk assessment for macro-materials is therefore insufficient to ensure their safety. There are different sources of nanoparticles; they exist in nature (from fires, volcanic eruptions, and nature structures) and can be produced by human activities under control or unintentionally from explosions, mechanical work, industrial processes, and others. The exposure levels have increased as manufactured nanomaterials become more widespread due to the pandemic, considering the wide applications like masks and disinfectant products with nanoparticles (Mosselhy et al. 2021, Rai et al. 2020). The most important production of nanoparticles takes place with the help of well-known chemical and engineering processes. There is enough evidence to suggest that exposure to nanoparticles, particularly those insoluble in water, should be minimized as a precaution. In the initial stages of nanotechnology, workers in nanotechnology research and nanotechnology companies were the main focus of exposure to manufactured nanoparticles. However, during the last years, more and more consumers have been exposed to manufactured nanoparticles, and at the same time, more nanoparticles can be potentially released into the environment. Most contemporary research on the impact of nanomaterials points towards synthetic or engineered systems, and there is a lack of studies centered on the effects of natural sources of nanomaterials. During the last decade, multiple works and reviews reported the properties of nanomaterials, their characteristic, properties, route of exposure, and interaction with human cells and the environment (Basinas et al. 2018, Cronin et al. 2020, Lewis et al. 2019, Schulte et al. 2019).

3.1 Human Health and Nanoparticles

Nanoparticles can enter the human body through various paths: inhalation, ingestion, via the skin, or disperse into the environment (De Matteis 2017). The same characteristics in size and reactivity that make nanoparticles so interesting lead to a new category of potentially toxic substances; due to their size, most nanoparticles will quickly enter the lungs and reach the alveoli (Hoet et al. 2004a). These inhaled particles can have two significant effects: induce inflammation in the respiratory tract, and transport through the bloodstream to other organs or tissues of the body (Bakand et al. 2012). Among the first concerns with nanoparticles was the asbestos comparison in terms of risks due to their fiber or needle-like shape with high biopersistence (Hoet et al. 2004b). Though the comparison seems reasonable, there is evidence that the nanoscale fibers tend to cluster together rather than exist as single fibers, possibly reducing exposure and potential for toxic effects (Maynard et al. 2004). Some respiratory and cardiovascular diseases can be caused by carbon black and other organic molecules, with harmful consequences. Diseases associated with air pollution containing nanoparticles can cause premature deaths of workers in the gas, coal, and asphalt industries (Armstrong et al. 2004).

Nanoparticles can also enter the human body via other routes like the upper nose, mouth, and intestines (Barua and Mitragotri 2014). Penetration via the skin has not been studied conclusively, but it mainly depends on the size and surface properties of the nanoparticles. The entrance of nanoparticles through the skin might facilitate immune-mediated responses, neurotoxicity, or the production of reactive molecules that could lead to cell damage (Nafisi and Maibach 2018). Nanoparticles of titanium dioxide, used in some sun protection products, do not penetrate the skin and remain on the outer layer (Gontier et al. 2008, Pflücker et al. 1999) but can cause unwanted oxidative damage to other organisms like fishes (Kaviyani et al. 2020).

Nanoparticles can also pass through different cell membranes of mammalians and be absorbed (Koziara et al. 2003). The absorption of different nanoparticles by cells has been observed (Oh and Park 2014); the adsorption rate depends on the nanoparticles' distribution, aggregation, and sedimentation. When endocytosis or phagocytosis occurs, the nanoparticles are stored in the cells and can cause toxic reactions. Their toxicity is related to their small size, large surface area, and ability to produce reactive oxygen species (Nel et al. 2006) and can cause inflammation and fibrosis in multicellular organisms, while in unicellular organisms, they have an anti-oxidant property and cytotoxicity (Oberdörster et al. 2005). Recently, Cronin and coworkers reported the effect of nanomaterials on the human immune system detailing a nanosafety review (Cronin et al. 2020).

One important consideration when evaluating the possible adverse effects of nanoparticle exposure is the potential transformation of these materials in any life cycle phase, from production to storage stages. For example, unprotected ZnO powders can double their size after four years exposed to air (Thurber et al. 2012), and some nanoparticles can have variations of physical or/and chemical parameters by the effects of different factors like (Nowack et al. 2012):

- Photochemical: Modification of the material's properties by the incidence of a particular wavelength. It has been proven that the interaction of TiO_2 NPs with sunlight produces the appearance of ROS in living organisms, causing an increase in toxicity (Abbas et al. 2020).

- Oxidation and reduction: Changes in the oxidation state depending on the medium and external conditions (pH, oxidizing and/or reducing agents, reagents, stabilizers). It has been demonstrated that the oxidation of nanoparticles of Ag^0 to Ag^+ increases the toxicity (Reed et al. 2016).

- Adsorption or desorption: Changes in the interaction of chemical forces can lead to the interaction or desorption from a certain matrix. Metal and organic NPs may adsorb on the surface of graphene oxide, affecting their availability and mobility and increasing their risk of toxicity (Abbas et al. 2020).

- Dissolution: Generation of ions or water-soluble molecules. For example, metallic nanoparticles of Cu or Zn can dissolve into ions, increasing their toxicity (Aruoja et al. 2009).

- Abrasion: Some erosive processes can produce NPs. The constant blackening of bicycle chains and rock-climbing ropes can release nanoparticles that may impact human health (Moosmüller et al. 2021).

- Biotransformation: Including processes mediated by biological agents. Selenium can be transformed into seleno-nanoparticles by lactic acid bacteria (Martínez et al. 2020).

There are no reports about the toxicity of nanomaterials under extreme conditions until now. With the production of engineered nanoparticles and their presence in more common functional objects, we are confronted with new materials that need more information describing the health risk of their potential transformations. This area is still evolving, and many questions are still open. The broad diversity of nanoparticles is being investigated, considering different parameters that influence the functionality and interaction with the human body and the environment. Biodegradable nanoparticles can be a promising alternative to reduce accumulation in the body, minimizing or even excluding long-term effects.

3.2 Environmental Risks

Besides risks related to human exposure, accidental releases of waste material of engineered nanoparticles into the environment can potentially occur. The possible routes for exposure to the environment range from discharge or leakage from final products, discharge from waste, the release of nanoparticles during the use of the products, diffusion, transport, and transformation in air, soil, and water. The bulk material properties are insufficient to classify the same material's environmental risks in the form of nanoparticles, as increased bioavailability is consistent with higher toxicities often found in nanosystems compared with their larger counterparts (Karlsson et al. 2009). Therefore, the possible environmental impact of each nanomaterial requires a detailed assessment of the characteristics of these systems, such as physicochemical properties, including how they are emitted to the environment and their toxicity in living beings (Dong et al. 2020).

The main criteria used to measure the environmental and human health risks are toxicity, persistence, and bioaccumulation (Thoeye et al. 2003). Highly toxic substances have high persistence and can concentrate in fatty tissues. The toxicity is conditioned by the ability of nanoparticles to reach and disperse into the different environmental compartments like air, earth, and water. Thus, the risk depends on how the nanoparticles are released into the environment (Rajkovic et al. 2020), and the quantification requires exhaustive research from the nanomaterials' production processes to the recycling and disposal (Nowack 2017), taking into account how they are incorporated into final products and how they are employed (Gottschalk et al. 2013).

One key point for the dispersion of nanoparticles into the environment is the aquatic media (Hartmann et al. 2014). The concentration of nanoparticles in surface water varies according to the type of material and the location. The European Union reports a high concentration of TiO_2 in surface water with a value of 2.2 $\mu g \cdot L^{-1}$ followed by silver nanoparticles with 1.5 $ng \cdot L^{-1}$ (Bundschuh et al. 2018, Sun et al. 2016). As stated, nanoparticles come from natural (soil erosion, dust storms, forest fires, and volcanic activities) and anthropogenic activities like drug production, groundwater remediation, burning fossil fuels, mining, and demolition (Singh 2015, Smita et al. 2012). In addition, nanoparticles can enter the aquatic environment through atmospheric release and water infiltration into the soil (Weinberg et al. 2011). Nanoparticles in water sources can transform depending on their intrinsic properties through physical, chemical, or biological transformations (Lowry et al. 2012). These transformations can affect their toxicity mainly by interacting with other pollutants or chemical conversion; for example, copper nanoparticles can be oxidized to Cu^{2+}, which is more toxic than the neutral form (Batley et al. 2013, Bielmyer et al. 2006). One of the first studies on the toxic effects of nanoparticles on aquatic organisms was published by Oberdörster in 2004. The largemouth bass was exposed to the uncoated fullerene C_{60}. Significant lipid peroxidation was reported in the brain of the animals after exposure, demonstrating the impact that engineered nanoparticles might have on aquatic and possibly other organisms (Oberdörster 2004). The concentration of nanoparticles in water must always be controlled, small concentrations (5–50 µg/L) may cause physiological changes, chromosomal alterations, and oxidative stress, but 1 mg/L can produce mortality (Ovissipour et al. 2013). A recent review summarizes the potential impact of nanoparticles in water environments, describing the effects on microbes, invertebrates, vertebrates, plants, and humans (Turan et al. 2019).

In the case of air pollution, there is vast experience in the particulate matter at the scale of µm. However, the information about the effects of anthropogenic nanoparticles is still under study. Ultrafine particulate matter (PM with a diameter less than 100 nm) is likely to have an even greater biological action than larger PM on an equivalent mass basis. The sources of most of these particles are industrial activity, diesel motors, or cars with defective catalysts. Translocation of nanoparticles into systemic circulation could underlie their toxicity to multiple organs (Raftis and Miller 2019). One crucial piece of evidence is the potential of nanoparticles to act as carriers for toxic chemicals on their surface in the human respiratory system (Zhiqiang et al. 2000). More studies are needed to understand the effects of nanoparticles present in the air to predict and avoid any probable toxicity of nanomaterials for human health and the environment.

Nanoparticles also impact soils; thus, it is essential to characterize their entry, accumulation, and migration. NPs can disperse into an environment through the direct outlet, for example, from wastewater treatment plants, or during their degradation, like loss of the protective cover. On the other hand, nanotechnology remediation is responsible for discharging a vital portion of nanoparticles into the environment. They can easily travel to groundwater and remain in the soils (Attia and Elsheery 2020, Sun 2019). Supplementary ways of environmental exposure include leakage within the production and transport steps of NPs or related products (Rajput et al. 2020, 2019). Nanoparticles can interact with plants' roots by being absorbed from the roots' surface and entering the cell wall. Depending on the size and type of nanoparticles, nanoparticles may enter the intercellular space, which the membranes can absorb. The surfaces of plants' cells can allow the entry of negatively-charged species into the intercellular space of the roots (Nowack et al. 2006). It has been proved that uncoated aluminum oxide nanoparticles harm the growth of plants' roots (Yang and Watts 2005). Still, titanium dioxide nanoparticles have a positive effect on the growth of some plants. Unlike larger samples of titanium dioxide, titanium dioxide nanoparticles can enhance enzymatic activity, increase the absorption of nitrates, and accelerate the conversion of inorganic nitrogen to organic nitrogen (Hund-Rinke and Simon 2006). Other studies on fishes and Daphnia have accounted for the accumulation of nanoparticles in the body, eggs, and inside the cells (Kashiwada 2006, Zhu et al. 2006). Carbon nanotubes have toxic effects on the respiratory systems of rainbow trout and increase the mortality in some freshwater crab species (Templeton et al. 2006).

One crucial issue that has caught the attention of researchers is nanoplastics. They are present in soils and water and are fractions from the fragmentation of larger plastic debris (Gigault et al. 2021) by heat, UV degradation, and biological factors, among others (Gigault et al. 2021). They can travel across biological membranes and access specific endocytosis pathways. Nanoplastics have more toxicity than larger-scale materials and have the potential to trigger reactive oxygen species production (Johnston et al. 2000). Incidental nanoparticles produced from the fragmentation of plastic waste are an essential element in the life cycle of plastic waste. More studies are necessary to understand the problem of plastic waste and its potential toxic effects on human and environmental health. The information about nanoparticles and the environment is still under study. Up to date, there is a lack of reports on how nanoparticles behave in different scenarios like air, water, or soil or their ability to accumulate in food chains. It is essential to mention that the complexity of the problem needs efforts from different scientific areas and governments. Basic information about behavior, mobility, and availability is required to assess their risks, added to accurate maximum concentrations.

4. Regulations

Nanomaterials possess specific and characteristic features that make them superior in catalysis, biomedical applications, remediation, etc. However, these qualities are also related to potential toxicity. Several governments and organizations have implemented regulations to minimize or avoid risks associated with nanomaterials. To date, there is no specific international regulation for production, handling, testing toxicity, and evaluating the environmental impact of nanoparticles. The US National Science Foundation issued the first broad evaluation of nanotechnology in 2001 (Bainbridge 2001). A few years later, the Office for Technology Evaluation at the German Bundestag TAB published a similar report (Paschen et al. 2003). In 2004, the UK Royal Society and Royal Academy of Engineering also strongly recommended using nanomaterials (Grobert and Hutton 2004). However, none of these reports systematically classified different nanotechnology risks and had limited access to experts.

After the first appearance of nanotechnology, there was high interest on the potential toxicity and impacts on human health. Medical standards were then modified to cover nanomaterials in the biomedical field (van Calster and D'Silva 2009). Since then, many regulations have been published in the USA and the European Union (EU) to control and decrease the potential risks of nanomaterials. According to EU Commission, nanomaterials are "a natural, incidental or manufactured material containing particles, in an unbound state or as an aggregate or as an agglomerate, and where for 50% or more of the particles in the number size distribution, one or more external dimensions is in size range of 1 nm to 100 nm", and the legislation guarantees that a nanomaterial in one sector will also be treated as such when it is used in another sector. In the EU, nanomaterials are regulated by the substance definitions of the European chemical agency (REACH) and the European Classification and Labelling of Chemicals (CLP) (Agency 2020). In addition, in 2009, the EU launched the Scientific Committee on Emerging and Newly Identified Health Risks (SCENIHR) to estimate risks associated with nanomaterials. In the USA, the Food and Drug Administration (FDA), the United States Environmental Protection Agency (USEPA), and the Institute for Food and Agricultural Standards (IFAS) have started different protocols about the risks of nanomaterials and how they can be avoided or diminished. Some non-governmental articles have also exposed the demands about nanomaterial regulations, health and safety of the public and workers, transparency, public participation, environmental protection, and the inclusion of broader impacts and manufacturer responsibility (Bowman 2010, Kimbrell 2009). Other public associations worldwide have also stated their concerns about the direct interaction of nanomaterials with the human body. This is the case with the Cosmetic Regulation Agencies in Canada and Japan, the organic suppliers, including the UK Soil Association, the Biological Farmers of Australia, and the Canada General Standards Board. In Latin America, some countries have made nanotechnology a development priority in their public policy platforms. Mexico released the first guidelines for the regulation of nanotechnology in 2012 (Foladori and Lau 2014), while Brazil also initiated actions to state a legal framework for the treatment and management of nanomaterials (Engelmann et al. 2018).

Governments, researchers, consumers, and manufacturers should be educated on nanomaterials' regulatory laws and risks. Some are not intrinsically hazardous and seem to be nontoxic, while others have beneficial health effects. However, considering all possible factors, a complete risk assessment is necessary to avoid negative impacts.

5. Conclusion

Although nanotechnology provides numerous benefits, there is currently concern about the unsafe disposal and massive use of various nanomaterials every year. The complex role of anthropogenic nanomaterials needs critical attention soon and actions from different world nations. Future research should focus on developing easy, economic, and highly reproducible analytical tools to understand the role of nanomaterials in a different environmental matrix. Understanding the interaction of nanomaterials in abiotic and biotic ecosystems will help formulate new regulations to mitigate the transmission and exposition of humans. Nanotoxicity studies should also play a central part in researchers' future agenda to look rigorously into natural nanomaterials' effect, translocation, and degradation inside the body. All these actions need the expertise of different research areas, so more collaborative opportunities are necessary to guide the development and application of sampling procedures, quantification, modeling approaches, and data curation techniques. Governments and regulatory agencies should then use this information to protect human health and the environment. Computational tools are a potential strategy for investigating and predicting the toxicological properties of various nanomaterials. Highly curated data available in the future will boost the development of computational nanotoxicology to shed light on the fundamental properties that affect nanomaterials' toxicity, offering safe-by-design criteria for a promising nanotechnological era.

Acknowledgments

M.A. thanks FONDECYT grant "Iniciación a la Investigación" #11180616, and Universidad Mayor grant "Proyecto Iniciación-2019091". M.B.C. thanks FONDECYT Regular #1180023.

References

Abbas, Q., Yousaf, B., Amina Ali, M.U., Munir, M.A.M., El-Naggar, A., ... Naushad, M. (2020). Transformation pathways and fate of engineered nanoparticles (ENPs) in distinct interactive environmental compartments: A review. *Environment International*, 138: 105646. doi:https://doi. org/10.1016/j.envint.2020.105646.

Agency, E.C. (2020). Nanomaterials. Retrieved from https://echa.europa.eu/regulations/nanomaterials.

Armstrong, B., Hutchinson, E., Unwin, J. and Fletcher, T. (2004). Lung cancer risk after exposure to polycyclic aromatic hydrocarbons: a review and meta-analysis. *Environmental Health Perspectives*, 112(9): 970–978. doi:doi:10.1289/ehp.6895.

Aruoja, V., Dubourguier, H.-C., Kasemets, K. and Kahru, A. (2009). Toxicity of nanoparticles of CuO, ZnO and TiO2 to microalgae Pseudokirchneriella subcapitata. *Science of the Total Environment*, 407(4): 1461–1468. doi:https://doi.org/10.1016/j.scitotenv.2008.10.053.

Attia, T.S. and Elsheery, N. (2020). Nanomaterials: Scope, applications, and challenges in agriculture and soil reclamation. In *Sustainable Agriculture Reviews* 41(pp. 1–39): Springer.

Bainbridge, W.S. (2001). *Societal Implications of Nanoscience and Nanotechnology*: Springer.

Bakand, S., Hayes, A. and Dechsakulthorn, F. (2012). Nanoparticles: A review of particle toxicology following inhalation exposure. *Inhalation Toxicology*, 24(2): 125–135. doi:10.3109/08958378.20 10.642021.

Barua, S. and Mitragotri, S. (2014). Challenges associated with penetration of nanoparticles across cell and tissue barriers: A review of current status and future prospects. *Nano today*, 9(2): 223–243. doi:10.1016/j.nantod.2014.04.008.

Basinas, I., Jiménez, A.S., Galea, K.S., Tongeren, M.v. and Hurley, F. (2018). A systematic review of the routes and forms of exposure to engineered nanomaterials. *Annals of Work Exposures and Health*, 62(6): 639–662. doi:10.1093/annweh/wxy048.

Batley, G.E., Kirby, J.K. and McLaughlin, M.J. (2013). Fate and risks of nanomaterials in aquatic and terrestrial environments. *Accounts of Chemical Research*, 46(3): 854–862.

Bielmyer, G.K., Grosell, M. and Brix, K.V. (2006). Toxicity of silver, zinc, copper, and nickel to the copepod acartia tonsa exposed via a phytoplankton diet. *Environmental Science and Technology*, 40(6): 2063–2068. doi:10.1021/es051589a.

Bowman, D.M. (2010). Global perspectives on the oversight of nanotechnologies. In *Nanotechnology Environmental Health and Safety* (pp. 73–95): Elsevier.

Brown, D. (2002). US regulators want to know whether nanotech can pollute. *Small Times*, 15.

Bundschuh, M., Filser, J., Lüderwald, S., McKee, M.S., Metreveli, G., Schaumann, G.E., ... Wagner, S. (2018). Nanoparticles in the environment: Where do we come from, where do we go to? *Environmental Sciences Europe*, 30(1): 1–17.

Cronin, J.G., Jones, N., Thornton, C.A., Jenkins, G.J.S., Doak, S.H. and Clift, M.J.D. (2020). Nanomaterials and innate immunity: A perspective of the current status in nanosafety. *Chemical Research in Toxicology*, 33(5): 1061–1073. doi:10.1021/acs.chemrestox.0c00051.

De Matteis, V. (2017). Exposure to inorganic nanoparticles: Routes of entry, immune response, biodistribution and *in vitro/in vivo* toxicity evaluation. *Toxics*, 5(4): 29. doi:10.3390/toxics5040029.

Dong, H., Li, L., Wang, Y., Ning, Q., Wang, B. and Zeng, G. (2020). Aging of zero-valent iron-based nanoparticles in aqueous environment and the consequent effects on their reactivity and toxicity. *Water Environment Research*, 92(5): 646–661. doi:https://doi.org/10.1002/wer.1265.

Eaglesham, D.J. (2005). The nano age? *MRS Bulletin*, 30(4): 260–261.

Edwards, B.C. (2000). Design and Deployment of a Space Elevator. *Acta Astronautica*, 47(10): 735–744. doi:https://doi.org/10.1016/S0094-5765(00)00111-9.

Engelmann, W., Hohendorff, R.V. and Leal, D.W.S. (2018). Nanotechnological regulations in Brazil. pp. 343–364. *In*: Rai, M. and Biswas, J.K. (eds.). *Nanomaterials: Ecotoxicity, Safety, and Public Perception*. Cham: Springer International Publishing.

Ezema, I.C., Ogbobe, P.O. and Omah, A.D. (2014). Initiatives and strategies for development of nanotechnology in nations: A lesson for Africa and other least developed countries. *Nanoscale Research Letters*, 9(1): 133–133. doi:10.1186/1556-276X-9-133.

Foladori, G. and Lau, E.Z. (2014). The regulation of nanotechnologies in Mexico. *Nanotech. L. and Bus.*, 11: 164.

Gigault, J., El Hadri, H., Nguyen, B., Grassl, B., Rowenczyk, L., Tufenkji, N., ... Wiesner, M. (2021). Nanoplastics are neither microplastics nor engineered nanoparticles. *Nature Nanotechnology*, 16(5): 501–507. doi:10.1038/s41565-021-00886-4.

Gontier, E., Ynsa, M.-D., Bíró, T., Hunyadi, J., Kiss, B., Gáspár, K., ... Surlève-Bazeille, J.-E. (2008). Is there penetration of titania nanoparticles in sunscreens through skin? A comparative electron and ion microscopy study. *Nanotoxicology*, 2(4): 218–231. doi:10.1080/17435390802538508.

Gottschalk, F., Sun, T. and Nowack, B. (2013). Environmental concentrations of engineered nanomaterials: Review of modeling and analytical studies. *Environmental Pollution*, 181: 287–300. doi:https://doi.org/10.1016/j.envpol.2013.06.003.

Grobert, N. and Hutton, D. (2004). Nanoscience and nanotechnologies: Opportunities and uncertainties. *London the Royal Society the Royal Academy of Engineering Report*, 46: 618–618.

Hartmann, N.I.B., Skjolding, L.M., Hansen, S.F., Baun, A., Kjølholt, J. and Gottschalk, F. (2014). Environmental fate and behaviour of nanomaterials: New knowledge on important transfomation processes. Danish Environmental Protection Agency. Environmental Project No. 1594.

Hoet, P.H., Brüske-Hohlfeld, I. and Salata, O.V. (2004b). Nanoparticles - known and unknown health risks. *Journal of Nanobiotechnology*, 2(1): 12–12. doi:10.1186/1477-3155-2-12.

Hoet, P.H.M., Nemmar, A. and Nemery, B. (2004a). Health impact of nanomaterials? *Nature Biotechnology*, 22(1): 19–19. doi:10.1038/nbt0104-19.

Hund-Rinke, K. and Simon, M. (2006). Ecotoxic effect of photocatalytic active nanoparticles (TiO2) on algae and daphnids (8 pp). *Environmental Science and Pollution Research*, 13(4): 225–232.

Johnston, C.J., Finkelstein, J.N., Mercer, P., Corson, N., Gelein, R. and Oberdörster, G. (2000). Pulmonary effects induced by ultrafine PTFE particles. *Toxicology and Applied Pharmacology*, 168(3): 208–215. doi:https://doi.org/10.1006/taap.2000.9037.

Karlsson, H.L., Gustafsson, J., Cronholm, P. and Möller, L. (2009). Size-dependent toxicity of metal oxide particles—A comparison between nano- and micrometer size. *Toxicology Letters*, 188(2): 112–118. doi:https://doi.org/10.1016/j.toxlet.2009.03.014.

Kashiwada, S. (2006). Distribution of nanoparticles in the see-through medaka (<i>Oryzias latipes</i>). *Environmental Health Perspectives*, 114(11): 1697–1702. doi:doi:10.1289/ehp.9209.

Kaviyani, F.E., Naeemi, A.S. and Salehzadeh, A. (2020). Acute toxicity and effects of titanium dioxide nanoparticles (TiO2 NPs) on some metabolic enzymes and hematological indices of the endangered Caspian trout juveniles (Salmo trutta caspius Kessler, 1877). *IFRO*, 19(3): 1253–1267. Retrieved from http://jifro.ir/article-1-2484-en.html.

Kimbrell, G.A. (2009). Governance of nanotechnology and nanomaterials: Principles, regulation, and renegotiating the social contract. *The Journal of Law, Medicine and Ethics*, 37(4): 706–723.

Klaine, S.J., Koelmans, A.A., Horne, N., Carley, S., Handy, R.D., Kapustka, L., ... von der Kammer, F. (2012). Paradigms to assess the environmental impact of manufactured nanomaterials. *Environmental Toxicology and Chemistry*, 31(1): 3–14. doi:https://doi.org/10.1002/etc.733.

Koshovets, O.B. and Ganichev, N.A. (2017). Nanotechnology and the new technological revolution: Expectations and reality. *Studies on Russian Economic Development*, 28(4): 391–397. doi:10.1134/S1075700717040104.

Koziara, J.M., Lockman, P.R., Allen, D.D. and Mumper, R.J. (2003). *In Situ* blood–brain barrier transport of nanoparticles. *Pharmaceutical Research*, 20(11): 1772–1778. doi:10.1023/B:PHAM.0000003374.58641.62.

Lewis, R.W., Bertsch, P.M. and McNear, D.H. (2019). Nanotoxicity of engineered nanomaterials (ENMs) to environmentally relevant beneficial soil bacteria—A critical review. *Nanotoxicology*, 13(3): 392–428. doi:10.1080/17435390.2018.1530391.

Lowry, G.V., Gregory, K.B., Apte, S.C. and Lead, J.R. (2012). Transformations of nanomaterials in the environment. *In*: ACS Publications.

Manoharan, M. (2008). Research on the frontiers of materials science: The impact of nanotechnology on new material development. *Technology in Society*, 30(3-4): 401–404.

Martínez, F.G., Moreno-Martin, G., Pescuma, M., Madrid-Albarrán, Y. and Mozzi, F. (2020). Biotransformation of selenium by lactic acid bacteria: formation of seleno-nanoparticles and seleno-amino acids. *Frontiers in Bioengineering and Biotechnology*, 8(506). doi:10.3389/fbioe.2020.00506.

Maynard, A.D., Baron, P.A., Foley, M., Shvedova, A.A., Kisin, E.R. and Castranova, V. (2004). Exposure to carbon nanotube material: Aerosol release during the handling of unrefined single-walled carbon nanotube material. *Journal of Toxicology and Environmental Health, Part A*, 67(1): 87–107. doi:10.1080/15287390490253688.

Miller, J.C., Serrato, R., Represas-Cardenas, J.M. and Kundahl, G. (2004). *The Handbook of Nanotechnology: Business, Policy, and Intellectual Property Law*: John Wiley and Sons.

Moosmüller, H., Giri, R., Sorensen, C.M. and Berg, M.J. (2021). Black metal nanoparticles from abrasion processes in everyday life: Bicycle drivetrains and rock-climbing ropes. *Optics Communications*, 479: 126413. doi:https://doi.org/10.1016/j.optcom.2020.126413.

Mosselhy, D.A., Virtanen, J., Kant, R., He, W., Elbahri, M. and Sironen, T. (2021). COVID-19 pandemic: What about the safety of anti-coronavirus nanoparticles? *Nanomaterials*, 11(3): 796.

Nafisi, S. and Maibach, H.I. (2018). Chapter 3—Skin Penetration of Nanoparticles. pp. 47–88. *In*: Shegokar, R. and Souto, E.B. (eds.). *Emerging Nanotechnologies in Immunology*. Boston: Elsevier.

Nel, A., Xia, T., Mädler, L. and Li, N. (2006). Toxic potential of materials at the nanolevel. *Science*, 311(5761): 622–627. doi:10.1126/science.1114397.

Nowack, B. (2017). Evaluation of environmental exposure models for engineered nanomaterials in a regulatory context. *NanoImpact*, 8: 38–47. doi:https://doi.org/10.1016/j.impact.2017.06.005.

Nowack, B., Ranville, J.F., Diamond, S., Gallego-Urrea, J.A., Metcalfe, C., Rose, J., ... Klaine, S.J. (2012). Potential scenarios for nanomaterial release and subsequent alteration in the environment. *Environmental Toxicology and Chemistry*, 31(1): 50–59. doi:https://doi.org/10.1002/etc.726.

Nowack, B., Schulin, R. and Robinson, B.H. (2006). Critical assessment of chelant-enhanced metal phytoextraction. *Environmental Science and Technology*, 40(17): 5225–5232.

Oberdörster, E. (2004). Manufactured nanomaterials (fullerenes, C60) induce oxidative stress in the brain of juvenile largemouth bass. *Environmental Health Perspectives*, 112(10): 1058–1062. doi:10.1289/ehp.7021.

Oberdörster, G., Oberdörster, E. and Oberdörster, J. (2005). Nanotoxicology: An emerging discipline evolving from studies of ultrafine particles. *Environmental Health Perspectives*, 113(7): 823–839. doi:10.1289/ehp.7339.

Oh, N. and Park, J.-H. (2014). Endocytosis and exocytosis of nanoparticles in mammalian cells. *International Journal of Nanomedicine*, 9 Suppl 1(Suppl 1): 51–63. doi:10.2147/IJN.S26592.

Ovissipour, M., Rasco, B. and Sablani, S. (2013). Impact of engineered nanoparticles on aquatic organisms. *J. Fisheries Livest. Prod.*, 1: e106.

Paschen, H., Coenen, C., Fleischer, T., Grünwald, R., Oertel, D. and Revermann, C. (2003). TA-Projekt.

Pflücker, F., Hohenberg, H., Hölzle, E., Will, T., Pfeiffer, S., Wepf, R., ... Gers-Barlag, H. (1999). The outermost stratum corneum layer is an effective barrier against dermal uptake of topically applied micronized titanium dioxide. *International Journal of Cosmetic Science*, 21(6): 399–411. doi:10.1023/A:1005450606035.

Raftis, J.B. and Miller, M.R. (2019). Nanoparticle translocation and multi-organ toxicity: A particularly small problem. *Nano Today*, 26: 8–12. doi:https://doi.org/10.1016/j.nantod.2019.03.010.

Rai, P.K., Usmani, Z., Thakur, V.K., Gupta, V.K. and Mishra, Y.K. (2020). Tackling COVID-19 pandemic through nanocoatings: Confront and exactitude. *Current Research in Green and Sustainable Chemistry*, 3: 100011.

Rajkovic, S., Bornhöft, N.A., van der Weijden, R., Nowack, B. and Adam, V. (2020). Dynamic probabilistic material flow analysis of engineered nanomaterials in European waste treatment systems. *Waste Management*, 113: 118–131. doi:https://doi.org/10.1016/j.wasman.2020.05.032.

Rajput, V., Minkina, T., Sushkova, S., Behal, A., Maksimov, A., Blicharska, E., ... Barsova, N. (2020). ZnO and CuO nanoparticles: a threat to soil organisms, plants, and human health. *Environmental Geochemistry and Health*, 42(1): 147–158. doi:10.1007/s10653-019-00317-3.

Rajput, V.D., Minkina, T., Sushkova, S., Chokheli, V. and Soldatov, M. (2019). Toxicity assessment of metal oxide nanoparticles on terrestrial plants. In *Comprehensive Analytical Chemistry* (Vol. 87, pp. 189–207): Elsevier.

Reed, R.B., Zaikova, T., Barber, A., Simonich, M., Lankone, R., Marco, M., ... Westerhoff, P.K. (2016). Potential environmental impacts and antimicrobial efficacy of silver- and nanosilver-containing textiles. *Environmental Science and Technology*, 50(7): 4018–4026. doi:10.1021/acs.est.5b06043.

Research, A.M. (2020). Nanotechnology Market By Type (Nanosensor and Nanodevice) and Application (Electronics, Energy, Chemical Manufacturing, Aerospace and Defense, Healthcare, and Others): Global Opportunity Analysis and Industry Forecast, 2021–2030. (2020). Retrieved from https://www.alliedmarketresearch.com/nanotechnology-market.

Roco, M.C. (2011). The long view of nanotechnology development: The National nanotechnology initiative at 10 years. *Journal of Nanoparticle Research*, 13(2): 427–445. doi:10.1007/s11051-010-0192-z.

Roco, M.C. and Bainbridge, W.S. (2005). Societal implications of nanoscience and nanotechnology: Maximizing human benefit. *Journal of Nanoparticle Research*, 7(1): 1–13. doi:10.1007/s11051-004-2336-5.

Saha, M. (2009). Nanomedicine: Promising tiny machine for the healthcare in future—A review. *Oman Medical Journal*, 24(4): 242–247. doi:10.5001/omj.2009.50.

Schulte, P.A., Leso, V., Niang, M. and Iavicoli, I. (2019). Current state of knowledge on the health effects of engineered nanomaterials in workers: A systematic review of human studies and epidemiological investigations. *Scandinavian Journal of Work, Environment and Health*, (3): 217–238. doi:10.5271/sjweh.3800.

Sengul, A.B. and Asmatulu, E. (2020). Toxicity of metal and metal oxide nanoparticles: A review. *Environmental Chemistry Letters*, 18(5): 1659–1683. doi:10.1007/s10311-020-01033-6.

Singh, A.K. (2015). *Engineered Nanoparticles: Structure, Properties and Mechanisms of Toxicity*: Academic Press.

Smita, S., Gupta, S.K., Bartonova, A., Dusinska, M., Gutleb, A.C. and Rahman, Q. (2012). Nanoparticles in the environment: Assessment using the causal diagram approach. *Environmental Health*, 11(1): S13. doi:10.1186/1476-069X-11-S1-S13.

Sun, H. (2019). Grand challenges in environmental nanotechnology. *Frontiers in Nanotechnology*, 1: 2.

Sun, T.Y., Bornhöft, N.A., Hungerbühler, K. and Nowack, B. (2016). Dynamic probabilistic modeling of environmental emissions of engineered nanomaterials. *Environmental Science and Technology*, 50(9): 4701–4711. doi:10.1021/acs.est.5b05828.

Taniguchi, N. (1974) On the Basic Concept of Nanotechnology. *Proceedings of the International Conference on Production Engineering*, Tokyo, 18–23.

Templeton, R.C., Ferguson, P.L., Washburn, K.M., Scrivens, W.A. and Chandler, G.T. (2006). Life-cycle effects of single-walled carbon nanotubes (SWNTs) on an estuarine meiobenthic copepod. *Environmental Science and Technology*, 40(23): 7387–7393.

Thoeye, C., Eyck, K., Bixio, D., Weemaes, M. and Gueldre, G. (2003). Methods used for health risk assessment. *State of the Art Report Health Risks in Aquifer Recharge using Reclaimed Water*. World Health Organization, Copenhagen, Denmark, 17–53.

Thurber, A.P., Alanko, G., II, G.L.B., Dodge, K.N., Hanna, C.B. and Punnoose, A. (2012). Unusual crystallite growth and modification of ferromagnetism due to aging in pure and doped ZnO nanoparticles. *Journal of Applied Physics*, 111(7): 07C319. doi:10.1063/1.3679147.

Turan, N.B., Erkan, H.S., Engin, G.O. and Bilgili, M.S. (2019). Nanoparticles in the aquatic environment: Usage, properties, transformation and toxicity—A review. *Process Safety and Environmental Protection*, 130: 238–249. doi:https://doi.org/10.1016/j.psep.2019.08.014.

van Calster, G. and D'Silva, J. (2009). Taking temperature—A review of european union regulation in nanomedicine. *European Journal of Health Law*, 16(3): 249–269. doi:https://doi.org/10.1163/157180909X453071.

Weinberg, H., Galyean, A. and Leopold, M. (2011). Evaluating engineered nanoparticles in natural waters. *TrAC - Trends in Analytical Chemistry*, 30(1): 72–83. doi:10.1016/j.trac.2010.09.006.

Yang, L. and Watts, D.J. (2005). Particle surface characteristics may play an important role in phytotoxicity of alumina nanoparticles. *Toxicology Letters*, 158(2): 122–132. doi:https://doi.org/10.1016/j.toxlet.2005.03.003.

Zhiqiang, Q., Siegmann, K., Keller, A., Matter, U., Scherrer, L. and Siegmann, H.C. (2000). Nanoparticle air pollution in major cities and its origin. *Atmospheric Environment*, 34(3): 443–451. doi:https://doi.org/10.1016/S1352-2310(99)00252-6.

Zhu, Y., Zhao, Q., Li, Y., Cai, X. and Li, W. (2006). The interaction and toxicity of multi-walled carbon nanotubes with Stylonychia mytilus. *Journal of Nanoscience and Nanotechnology*, 6(5): 1357–1364.

Index

For Product Safety Concerns and Information please contact our EU
representative GPSR@taylorandfrancis.com
Taylor & Francis Verlag GmbH, Kaufingerstraße 24, 80331 München, Germany